AF618968

Springer Oceanography

The Springer Oceanography series seeks to publish a broad portfolio of scientific books, aiming at researchers, students, and everyone interested in marine sciences. The series includes peer-reviewed monographs, edited volumes, textbooks, and conference proceedings. It covers the entire area of oceanography including, but not limited to, Coastal Sciences, Biological/Chemical/Geological/Physical Oceanography, Paleoceanography, and related subjects.

More information about this series at http://www.springer.com/series/10175

Bernd Schneider · Jens Daniel Müller

Biogeochemical Transformations in the Baltic Sea

Observations Through Carbon Dioxide Glasses

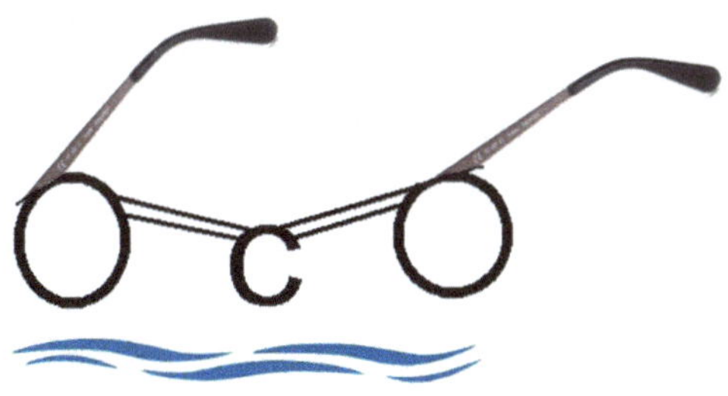

Springer

Bernd Schneider
Department of Marine Chemistry
Leibniz Institute for Baltic Sea Research
Warnemünde
Germany

Jens Daniel Müller
Department of Marine Chemistry
Leibniz Institute for Baltic Sea Research
Warnemünde
Germany

ISSN 2365-7677 ISSN 2365-7685 (electronic)
Springer Oceanography
ISBN 978-3-319-61698-8 ISBN 978-3-319-61699-5 (eBook)
DOI 10.1007/978-3-319-61699-5

Library of Congress Control Number: 2017945697

Printed on acid-free paper

This Springer imprint is published by Springer Nature
The registered company is Springer International Publishing AG
The registered company address is: Gewerbestrasse 11, 6330 Cham, Switzerland

Preface

This book is the product of our joint efforts to present the biogeochemistry of the Baltic Sea from a unique perspective, "through carbon dioxide glasses." We are convinced that this view advances our understanding of the organic matter production and mineralization processes that constitute the basic biogeochemical transformations in the sea and that trigger many secondary processes. The term "understanding" here means to be able to describe transformations as a function of time, to identify the controlling variables and finally to deduce the kinetics of the considered process. This allows us to explain the current biogeochemical status of the Baltic Sea but, more important, is a requisite for the use of numerical models to realistically predict the development of the ecological status of the Baltic Sea in a changing environment. Current biogeochemical models using state-of-the-art process parameterizations have increased our comprehension of the functioning of the Baltic Sea ecosystem and of the complex relationship between biogeochemistry and hydrography. Nonetheless, model simulations of the carbon cycle and thus of the marine CO_2 system either have failed to simulate the seasonal and regional patterns of the CO_2 system or were able to reproduce them only by invoking biogeochemical assumptions yet to be verified by observations. Therefore, investigations of the marine CO_2 system, which is intimately connected to all biogeochemical processes, have two functions: Identification of gaps in our process understanding and filling the gaps by improving the process parameterizations.

A recognition of the importance of studying the CO_2 system in the Baltic Sea developed during the past 25 years of related research carried out at the Leibniz Institute for Baltic Sea Research. The accumulation of a huge amount of data during 14 years of automated CO_2 measurements made from a cargo ship has provided the basis for several publications. This refers also to studies of the deepwater CO_2 accumulation in connection with the Institute's long-term observation programme. Here we present a synoptic view of the collected data and the gain in knowledge by connecting the results of previously published process studies to an overall picture. We consider our book as a contribution to the biogeochemical discussions within the Baltic Earth Network which is a scientific network for regional Earth system science for the Baltic Sea region (www.baltic.earth). We also hope that it will guide the inclusion of the carbon cycle into biogeochemical models and the integration of CO_2 measurements into Baltic Sea monitoring programs, such as proposed by the BONUS project INTEGRAL.

The cooperation between the authors in writing this book is also worthy of mention. Whereas BS was able to draw extensively on his own decades-long research in different fields of chemical oceanography, JDM is a Ph.D. student and thus at the beginning of his scientific career. His skills in dealing with large amounts of data and in the creative visualization of the inherent information were essential to the success of the investigations and the presentation of their results. The bridge between our generations enabled many fruitful and inspiring discussions from which both the book and we profited.

Regarding the automated surface CO_2 measurements on a cargo ship, we are much obliged to the Finnish Environmental Institute (SYKE), which has been using the cargo ship since 1993 for automated fluorescence (chlorophyll *a*) measurements and for taking samples for the analysis of nutrient concentrations within the Algaline Project. Due to the institute's generous cooperation, we were able to make use of the same infrastructure for the deployment of our own measurement system. Furthermore, our Finnish partners made nutrient data available for CO_2 data analyses and for use in this book. During 2012–2016, the operation of our CO_2 measurement system was generously funded by the German Federal Ministry of Education and Research (BMBF) in the frame of the German contribution to ICOS (Integrated Carbon Observation System) (Grant Numbers 01LK1101F and 01LK1224D).

We also thank the Swedish Agency for Marine and Water Management and the SMHI (Swedish Meteorological and Hydrographical Institute) for ready access to their monitoring database.

Finally, we sincerely appreciate the willingness of the following experts and colleagues to check the content of the book for scientific correctness and clarity: Profs. L. Anderson and D. Turner (University of Gothenburg), Prof. A. Körtzinger (Helmholtz Centre for Ocean Reserach Kiel, GEOMAR), and Dr. N. Wasmund and Dr. T. Seifert (Leibniz Institute for Baltic Sea Research, Warnemünde). Their thorough review resulted in a number of comments and recommendations that led to considerable improvements in the text and its presentation.

At the IOW, we are indebted to many of our colleagues, especially B. Sadkowiak, who constructed the automated measurement system's hardware and ensured proper operation of the system. We are also grateful to the Monitoring team for their regular collection of samples for total CO_2 investigations, and to the staff of the IOW's CO_2 lab for performing the respective analyses.

Warnemünde, Germany

Bernd Schneider
Jens Daniel Müller

Contents

About the Authors

Bernd Schneider studied Chemistry at the Christian-Albrechts-Universität in Kiel, Germany. His scientific career as a marine chemist started in 1975 as a Ph.D. student at the Institute of Marine Research in Kiel. In the years that followed he acted as a researcher at different research institutes in Germany and temporarily also at research facilities in the US and Sweden. For more than a decade he was working on the deposition of atmospheric trace substances before starting studies on the marine CO_2 system and the relationship to biogeochemical processes. Currently he is a senior scientist at the Leibniz Institute for Baltic Sea Research in Warnemünde, Germany. Photo: © Kristin Beck, IOW.

Jens Daniel Müller finished his Bachelors studies in Chemistry in 2009, whereafter he completed a Masters program in Biological Oceanography at Geomar in Kiel. Since 2014 he has been a Ph.D. student in Gregor Rehder's working group "Biogeochemistry of environmentally relevant gases". Within the framework of the BONUS project PINBAL (Grant No. 03F0689A) his focus is on spectrophotometric pH measurement in brackish waters. Besides his fascination for biogeochemical and CO_2 system research, he is an enthusiastic scientific diver and sailor. Photo: © Kristin Beck, IOW.

Abbreviations

Variables and Constants

[…]	Concentration, equivalent to c
$[H^+]$	Concentration of protons
A_T	Total alkalinity
c	Concentration, equivalent to []
C_T	Total inorganic carbon or total CO_2 (synonymous to DIC)
C_T*	Total inorganic carbon calculated from pCO_2 based on fixed S and A_T assumptions
DIC	Dissolved inorganic carbon (synonymous to C_T)
DOC	Dissolved organic carbon
F	Flux
fCO_2	CO_2 fugacity
f_D	Delay factor
$iNCP_t$	Depth integrated NCP_t
K_a	First dissociation constant of carbonic acid
K_0	Solubility constant of CO_2*
K_1	First dissociation constant of the marine CO_2 system
K_2	Second dissociation constant of the marine CO_2 system
K_S	Dissociation constant of hydrogen sulfate
K_B	Dissociation constant of boric acid
K_W	Ion product of water
K_{hyd}	Hydration constant of CO_2
k	Transfer velocity
k_{mix}	Mixing coefficient
K_{sp}	Solubility product of $CaCO_3$
NCP	Net community production of POC
NCP_t	Total NCP, including POC and DOC
OM	Organic matter
pCO_2	CO_2 partial pressure

pH Negative decadal logarithm of $[H^+]$ (subscripts: T for total scale, F for free scale, SWS for seawater scale, NBS for NBS scale)
POC Particulate organic carbon
POM Particulate organic matter
R Revelle factor
R* Modified Revelle factor
S Salinity
Sc Schmidt number
SST Sea surface temperature
T Temperature
t Time
u Wind speed
z_{eff} Effective mixing depth
z_{mix} Mixing depth
z_{pro} Maximum depth of OM production
τ Equilibration time
Ω Saturation of calcium carbonate minerals

Others

ARK Arkona Basin
BP Baltic Proper
EGS Eastern Gotland Sea
GB Gulf of Bothnia
GF Gulf of Finland
GR Gulf of Riga
HGF Helsinki Gulf of Finland
IOW Leibniz Institute for Baltic Sea Research Warnemünde
KA Kattegat Area
MEB Mecklenburg Bight
NGS Northern Gotland Sea
SMHI Swedish Meteorological and Hydrographical Institute
SYKE Finnish Environment Institute
VOS Voluntary Observing Ship
WGF Western Gulf of Finland
WGS Western Gotland Sea

Chapter 1
Introduction

1.1 History of CO_2 System Research in the Baltic Sea

Studies of the marine CO_2 system, based on measurements or modeling, have been the focus of chemical oceanography for roughly 30 years. This is because the oceans participate in the global CO_2 budget and have the potential to absorb a large fraction of anthropogenic CO_2 emissions. The Baltic Sea contributes only $\sim$0.002% to the volume of the world's oceans and thus plays only a minor role in this process. Nonetheless, the first fundamental experimental and theoretical studies on the marine CO_2 system were performed in the Baltic Sea. The pioneering work by Kurt Buch and others at the beginning of the 20th century provided chemical-analytical insights and the physico-chemical basis for a quantitative assessment of the marine CO_2 system. Their theoretical concepts and thermodynamic characterization of the CO_2 system as the major component of the marine acid-base system was still state of the art in the fifties and sixties of the last century. Furthermore, the first comprehensive field studies concerning the horizontal and vertical distribution of the characteristic variables of the CO_2 system, such as alkalinity (A_T), total CO_2 (C_T), pH, the partial pressure of CO_2 (pCO_2), and calcium carbonate saturation (Ω_{CaCo_3}), were performed between 1927 and 1936. The results allowed the mapping of the northern and central Baltic Sea with respect to these parameters (Buch 1945). An example for the outcome of this historic studies is given in Fig. 1.1, which shows the surface water pCO_2 distribution in the northern Baltic Sea during spring and autumn.

These investigations were stopped with the beginning of the Second World War and not continued thereafter. Only in the course of the development of Baltic Sea environmental monitoring efforts, pH and alkalinity were included as standard variables into the measurement program in some countries. However, the inherent biogeochemical information that emerged from these studies was not examined further. It took until the mid-1990s to revive research on the marine CO_2 system of the Baltic Sea, with the Leibniz Institute for Baltic Sea Research (IOW) in

B. Schneider and J.D. Müller, *Biogeochemical Transformations in the Baltic Sea*, Springer Oceanography, DOI 10.1007/978-3-319-61699-5_1

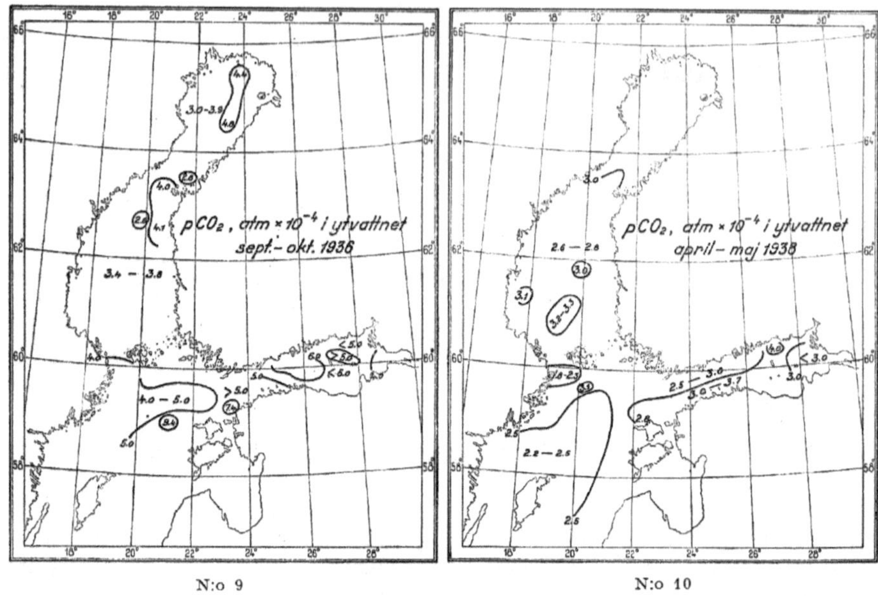

Fig. 1.1 Distribution of the CO_2 partial pressure in surface waters of the northern Baltic Sea as derived from measurements of alkalinity and pH during the first decades of the last century (Buch 1945)

Warnemünde playing a prominent role. Observations of the marine CO_2 system enabled biogeochemical process studies, including those on organic matter (OM) production and mineralization. Other issues such as the air-sea CO_2 gas exchange balance and the contribution to an overall carbon budget (Kulinski and Pempkowiak 2012) for the Baltic Sea played only a marginal role although not entirely ignored. Although several research cruises were performed, the knowledge about the seasonal, interannual, and regional variability of the CO_2 system remained fragmentary. A significant breakthrough was achieved in 2003, with the operation of a fully automated pCO_2 measurement system, mounted on a voluntary observation ship (VOS). Since then, data allowing characterization of the CO_2 system at high temporal and spatial resolution have been available for the entire Baltic Proper and constitute the data base for surface water biogeochemical studies presented in this book. Additionally, investigations of the deep-water CO_2 system were intensified since 2003 by including analyses of C_T into the long-term observation program of the IOW. C_T is determined at intervals of 2–3 months to obtain a record of CO_2 accumulation during stagnant water conditions, and thus facilitate investigations of the OM mineralization process. The pCO_2 and C_T data used in the following sections to analyze key biogeochemical processes in the Baltic Sea originated entirely from investigations performed at the IOW. At the time that this section was written, almost no other published pCO_2 and C_T data were available. The pCO_2 data from the IOW are partly openly available through the CDIAC and SOCAT data pools (http://cdiac.ornl.gov/oceans/coastal/baltic/.html).

1.2 Objectives of This Book

Many recent publications on the physico-chemistry of the marine CO_2 system and its importance for the global carbon budget emphasize anthropogenic perturbations, especially increasing CO_2 emissions. Our approach in this book differs in two aspects. First, by focusing on the Baltic Sea, with its biogeochemical and hydrographic peculiarities, we provide a clearly regional focus. Second, using observations of the temporal and spatial variability of the marine CO_2 system we identify and quantify biogeochemical processes, almost all of which are triggered by the formation (photosynthesis) and mineralization (respiration) of OM and are thus connected with the consumption and production of CO_2, respectively.

Before presenting the advantages of CO_2 studies for deepening our understanding of the processes that control the net community production (NCP) and the OM mineralization, we briefly recall the traditional methods for the determination of the NCP and OM mineralization. Estimates of the NCP are frequently based on measurements of either nutrient depletion or oxygen production in connection with the chemical bulk formula for photosynthesis (Eq. 1.1):

$$\begin{aligned} &106CO_2 + 16NO_3^- + HPO_4^{2-} + 122H_2O + 18H^+ \\ &\Rightarrow (CH_2O)_{106}(NH_3)_{16}H_3PO_4 + 138O_2 \end{aligned} \quad (1.1)$$

This method relies upon the stoichiometric coefficients in Eq. 1.1, which reflect the Redfield ratios (Redfield et al. 1963) for the elemental composition (C/N/P) of particulate organic matter (POM). This is especially critical with respect to the C/P ratio of POM, which may deviate by a factor of 4 from the Redfield ratio during the mid-summer production that is fueled by N-fixation. The C/N ratios of POM may also deviate from the Redfield ratio but the differences are much smaller. According to long-term monitoring data (IOW, 1995–2012) the mean ($\pm$SD) C/N ratio of POM in the upper 10 m of the Baltic Sea was 8.0($\pm$1.0) during the productive period from April to August, which is only $\sim$20% larger than the Redfield ratio of 6.6. However, nitrate consumption can be used to estimate OM production only during the first phase of the spring bloom, because nitrate is almost entirely exhausted by mid-April and any further net production is based on other N sources, such as N-fixation. Attempts were made to use oxygen production during photosynthesis to calculate OM production (Stigebrandt 1991). However, this approach is also problematic because the stoichiometric coefficient of 138 is based on the questionable assumption that the OM produced consists entirely of carbohydrates (Anderson 2000). Furthermore, oxygen released during photosynthesis may escape relatively quickly into the atmosphere via gas exchange and calculations to account for this effect are associated with large uncertainties.

These considerations reveal a number of shortcomings of traditional methods for the determination of the NCP which can be largely avoided using investigations of the marine CO_2 system. This approach does not depend on any questionable stoichiometric coefficients because the formation of any organic molecule requires an equivalent of CO_2 molecules. Furthermore, since re-equilibration with atmospheric

CO_2 is a slow process, the biological draw-down of total CO_2 and thus of pCO_2 is conserved for a relatively long time. Hence, the gas-exchange term and the associated uncertainties become less important in calculations of the loss of CO_2 by NCP. The use of pCO_2 measurements for estimates of the OM production is in particular suited for the Baltic Sea because calcifying plankton is virtually absent in the entire Baltic Sea except in the Kattegat region (Tyrrell et al. 2008). Hence, internal changes of alkalinity and total CO_2 (see Sect. 5.1.4) by the formation of calcium carbonate can be neglected and uncertainties associated with assumptions about the abundance of calcifyers can be avoided.

The mineralization of POM has been described and quantified as the loss of oxygen or as an increase of H_2S which reflects the use of sulfate as the oxidant. This approach is reasonable as it addresses the immediate consequences of mineralization. However, in studies aimed at determining the dynamics and stoichiometry of mineralization, measurements of O_2 and hydrogene sulphide (H_2S) are inadequate because they are not the only oxidants used in the mineralization process. Rather, manganese dioxide and nitrate may also contribute although a quantification of their contributions is difficult. Furthermore, H_2S may be removed from the water column by the precipitation of metal sulfides such as iron sulfide. Finally, also the oxidant consumption for OM mineralization depends on the composition of the OM (Anderson 2000). By contrast, the release and accumulation of C_T can be attributed exclusively to OM mineralization, provided that dissolution of carbonates does not occur. The latter assumption is justified because the input of calcium carbonate derived from calcifying plankton is negligible (see above). Vice versa, CO_2 released by mineralization is also not removed from the water column by the formation of calcium carbonates because the deep water is highly undersaturated with respect to calcium carbonate.

The results presented herein draw extensively on our own research at the IOW and have mostly been published in scientific journals during the past 15 years. By bringing together the insights derived from individual scientific achievements, we construct a coherent picture both of the seasonality and regional peculiarities of biomass production in the Baltic Sea and of the processes connected with OM mineralization in its deeper water layers. Our discussion also presents a strong case for including the marine CO_2 system into ecosystem or biogeochemical models. Early model simulations failed at reproducing the observed seasonality of the surface-water pCO_2, indicating fundamental shortcomings in our understanding of the factors that control biomass production in the central Baltic Sea. Since biological activity is directly related to changes in the CO_2 concentration, the ability of an ecosystem/biogeochemical model to simulate the seasonality of the marine CO_2 system is an essential criterion to ensure the quality of the model. The close coupling of biomass production and CO_2 loss also suggests the need to include data on the marine CO_2 system in monitoring eutrophication of the Baltic Sea. Eutrophication is defined as "an increase in the rate of supply of OM to an ecosystem," in which the term "supply" includes internal OM production (Nixon 1995). Accordingly, analyses of the CO_2 system are more informative than any other method, including traditional determinations of chlorophyll or nutrient concentrations, in detecting eutrophication.

The book is sub-divided into seven major chapters. Following the Introduction (this chapter), the physico-chemical fundamentals of the marine CO_2 system are discussed (Chap. 2). After a brief description of the general principles, we present the peculiarities of the Baltic Sea CO_2 system. These are mainly due to the complex relationship between the wide salinity range and the alkalinity distribution which affect the responses of the CO_2 system to biogeochemical transformations. Chapter 3 is a short summary of the most important hydrographic features of the Baltic Sea Proper that contribute to the control of the major biogeochemical processes: the production and mineralization of OM. We address especially the seasonality of the surface-water stratification and the interplay between the development of stagnant conditions and water-renewal events in the deep basins of the Baltic Sea Proper. The database used for our analysis of biogeochemical processes is presented in Chap. 4, which also includes a short summary of the CO_2 measurements conducted at the IOW during the last 20 years. Chapter 5 provides an analysis of the biogeochemical processes in the Baltic Sea on the basis of the long-term high-resolution pCO_2 data obtained from the VOS-based automated pCO_2 measurement system. We start with a holistic look at the spatio-temporal pCO_2 patterns for the 2003–2015 observation period. Thereafter, we demonstrate the potential of the data in assessing the control and extent of plankton bloom events, using data from the year 2009 as an example and taking into account the results of previously published data analysis. Chapter 6 shows the development of the deep water C_T concentrations during a period of ten years without any major water renewal event. We present OM mineralization rates and their relationship to the release of phosphate during a stagnation period that lasted from 2004 to 2014. Chapter 7 summarizes the gain in biogeochemical process understanding that was obtained by investigations of the marine CO_2 system, and the advantages in using the marine CO_2 system for biogeochemical studies, but we will also address unavoidable limitations of this approach.

References

Anderson LA (2000) On the hydrogen and oxygen content of marine organic matter. Deep Sea Res 42:1675–1680

Buch K (1945) Kolsyrejämvikten i Baltiska Havet. Fennia 68

Kulinski K, Pempkowiak J (2012) Carbon Cycling in the Baltic Sea. Geoplanet: Earth and Planetary Sciences, Springer Berlin Heidelberg. doi:10.1007/978-3-642-19388-0_1

Nixon SW (1995) Coastal marine eutrophication: a definition, social causes, and future concerns. Ophelia 41:199–219

Redfield AC, Ketchum BH, Richards FA (1963) The influence of organisms on the composition of sea water. In: Hill MN (Ed) The Sea, vol 2. Interscience, New York, pp 26–77

Stigebrandt A (1991) Computations of oxygen fluxes through the sea surface and the net production of organic matter with application to the Baltic and adjacent seas. Limnol Oceanogr 36(3):444–454

Tyrrell T, Schneider B, Charalampopoulou A, Riebesell U (2008) Coccolithophores and calcite saturation state in the Baltic and Black Seas. Biogeosciences 5:1–10

Chapter 2
The Marine CO_2 System and Its Peculiarities in the Baltic Sea

2.1 Atmospheric CO_2 Over the Baltic Sea

Due to its solubility and chemistry in aqueous solutions, CO_2 is distributed in a characteristic manner between the atmosphere and ocean waters. CO_2 inventories in these two compartments are coupled by gas exchange and are therefore mutually dependent. This is highly important for the atmospheric CO_2 budget, which has been strongly perturbed by anthropogenic CO_2 emissions. From the CO_2 released into the atmosphere mainly by the combustion of fossil fuel during the past ~200 years, ~30% has been taken up by the world's oceans (Le Quéré et al. 2016). This has considerably mitigated the increase in atmospheric CO_2 concentrations, which are currently at ~400 ppm (https://www.esrl.noaa.gov/gmd/ccgg/trends/). Although in the south the Baltic Sea borders heavily urbanized and industrialized areas, the atmospheric CO_2 concentration differs only slightly from those determined at background stations in the northern hemisphere. An example is given in Fig. 2.1, which shows the CO_2 concentrations in the atmosphere over the Baltic Sea in 2005 (Schneider et al. 2014). On average, they deviate only by 3 ppm from those at the remote station in Barrow, Alaska. Likewise, the seasonal peak-to-peak amplitudes of the atmospheric CO_2 concentrations (~17 ppm) caused by the uptake and release of CO_2 by the terrestrial biosphere do not show significant regional differences. The low impact of regional CO_2 sources on the atmospheric CO_2 level is due to rapid atmospheric mixing and the relatively long residence time of CO_2 in the atmosphere. The latter implies high mean concentrations and less influence of local or regional sources.

B. Schneider and J.D. Müller, *Biogeochemical Transformations in the Baltic Sea*, Springer Oceanography, DOI 10.1007/978-3-319-61699-5_2

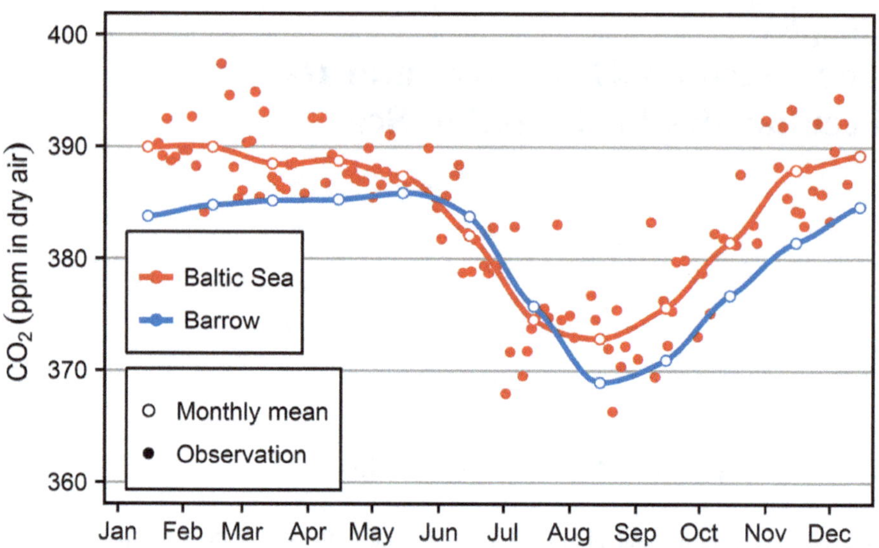

Fig. 2.1 Atmospheric CO_2 concentrations in the Baltic Sea area (58°N, 20°E, *red*) and at the northern hemispheric background station, Barrow, Alaska (71°N, 156°W, *blue*) in 2005. Only small differences with respect to the mean values and the seasonal amplitude exist

2.2 Aqueous Equilibrium Chemistry of CO_2

Once CO_2 is physically dissolved in water, it undergoes hydration and forms carbonic acid (Eq. 2.1):

$$CO_2 + H_2O \Leftrightarrow H_2CO_3 \tag{2.1}$$

This reaction is at equilibrium when the CO_2 and H_2CO_3 concentrations satisfy Eq. 2.2:

$$\frac{[H_2CO_3]}{[CO_2]} = K_{hyd} \tag{2.2}$$

The equilibrium between dissolved CO_2 and H_2CO_3 is thus characterized by the hydration constant K_{hyd}.

This is the first in a series of equilibrium reactions that control the abundance of CO_2 species in seawater. The chemical equilibria of this and all other chemical reactions concerning the marine CO_2 system are expressed in terms of concentrations rather than activities (Box 2.1). This convention applies to all CO_2 equilibria described in the following sections and all concentrations regarding CO_2 system variables are expressed as moles per mass seawater.

Box 2.1: Activities and concentrations

The activity (a) of a dissolved species is related to its concentration (c) by the activity coefficient γ:

$$a = \gamma \cdot c \tag{B2.1}$$

which accounts for the change in the chemical potential of a chemical species due to its interactions with any other solute. Classical thermodynamic equilibrium constants are based on activities and depend on temperature and pressure. However, equilibrium constants describing the marine CO_2 system are generally expressed in terms of concentration, which is more easily accessible by chemical analysis. Hence, a concentration-based equilibrium constant implicitly includes the corresponding activity coefficient. This implies that the equilibrium constants describing the marine CO_2 system depend also on salinity, which represents the major solute in seawater and is thus controlling the activity coefficient. This convention, together with the use of the concentration in units of moles per mass seawater, applies to all CO_2 equilibria described in the following sections.

The CO_2 hydration constant K_{hyd} (S = 35, T = 25 °C) is 1.2×10^{-3} (Soli and Byrne 2002) and indicates that only about one per mille of dissolved CO_2 exist as undissociated carbonic acid. Since H_2CO_3 molecules do not have a specific function in biogeochemical transformations, they are not explicitly included in the characterization of the marine CO_2 system; instead, the sum of the CO_2 and H_2CO_3 concentrations is used, expressed as the variable CO_2^* (synonymous to $H_2CO_3{}^*$) (Eq. 2.3):

$$[CO_2^*] = [CO_2] + [H_2CO_3] \tag{2.3}$$

which thus describes the solubility of CO_2 in seawater. The equilibrium between CO_2 in the gas phase and CO_2* dissolved in seawater is characterized by the solubility constant, K_0 (Eq. 2.4), which depends on temperature, pressure, and salinity (Weiss 1974). The pressure dependency of K_0 refers to the total pressure that acts upon the considered water body and is a function of water depth (Eq. 2.4).

$$[CO_2^*] = K_0 \cdot fCO_2 \tag{2.4}$$

where fCO_2 is the fugacity, expressed in pressure units. It determines the chemical potential of CO_2 in the gas phase and is thus analogous to the activity of dissolved species. It deviates slightly from the CO_2 partial pressure, pCO_2, due to the non-ideal behavior of CO_2. However, because the pCO_2 is directly accessible by measurements, our considerations of the CO_2 system will consistently use pCO_2, which for computations concerning the marine CO_2 system is then converted by the currently available software to fCO_2 (e.g., Pierrot et al. 2006; Lavigne et al. 2011).

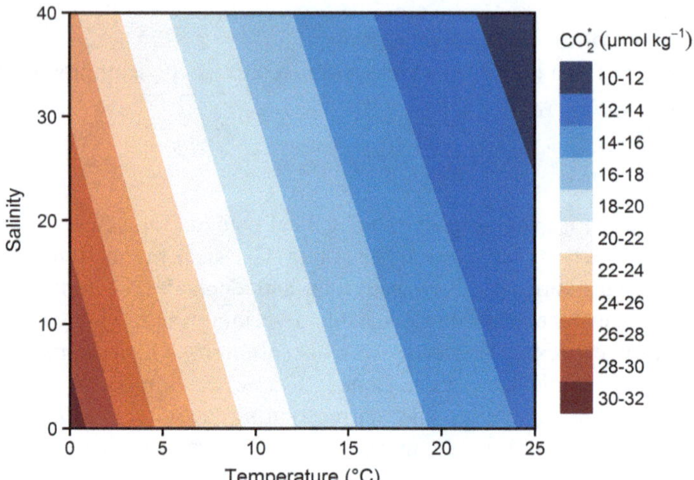

Fig. 2.2 CO_2^* concentrations of seawater at equilibrium with an atmospheric pCO_2 of 400 µatm. The pattern reflects the salinity and temperature dependence of the CO_2 solubility constant K_0

To illustrate variations in the solubility constant, K_0, the concentrations of CO_2^* at equilibrium with an atmospheric pCO_2 = 400 µatm are shown as a function of temperature and salinity in Fig. 2.2. The solubility constant and thus CO_2^* concentrations decrease with increasing temperature and salinity. The concentrations span a wide range and are as low as 12 µmol kg^{-1} at an oceanic salinity of 35 and a temperature of 25 °C but may be as high as >30 µmol kg^{-1} in river water with a salinity of ~0 and a temperature of 0 °C.

Carbonic acid produced by the hydration of CO_2 dissociates in a first step according to Eq. 2.5:

$$H_2CO_3 \Leftrightarrow HCO_3^- + H^+ \tag{2.5}$$

During the dissociation of an acid, hydrogen ions (protons) are transferred to a water molecule or to a cluster of several water molecules. Although free protons do not exist, it is common practice to denote combinations of protons and water molecules (e.g., H_3O^+) by H^+. The equilibrium condition for the first dissociation step of H_2CO_3 is given by Eq. 2.6:

$$\frac{[H^+] \cdot [HCO_3^-]}{[H_2CO_3]} = K_a \tag{2.6}$$

The dissociation constant, K_a, at 25 °C and 0 salinity is 2.5×10^{-4} mol kg^{-1}. This means that carbonic acid is a relatively strong acid, with a pK_a similar to that of formic acid. However, the acidity (pH) of carbonic acid dissolved in pure water is constrained by the equilibrium with dissolved CO_2 (Eqs. 2.1 and 2.2), which in turn is limited by the equilibrium with gas-phase CO_2. Only a small fraction of the dissolved CO_2 is transferred to H_2CO_3 and can thus act as an acid. It is therefore convenient to combine the hydration and the first dissociation equilibria and to describe the acidity of carbonic acid as a function of the dissolved CO_2 concentration, as shown in Eq. 2.7:

$$CO_2 + H_2O \Leftrightarrow [HCO_3^-] + [H^+] \tag{2.7}$$

Taking into account the CO_2 hydration equilibrium (Eq. 2.2), the first dissociation equilibrium of H_2CO_3 (Eq. 2.6), and the definition of CO_2^* (Eq. 2.3) yields the equilibrium condition for the reaction shown in Eq. 2.7 (see Box 2.2 for the derivation):

$$\frac{[H^+] \cdot [HCO_3^-]}{[CO_2^*]} = K_1 \tag{2.8}$$

with:

$$K_1 = K_a \cdot \frac{K_{hyd}}{1 + K_{hyd}} \tag{2.9}$$

where K_1 is the first dissociation constant of the marine CO_2 system (Eqs. 2.8 and 2.9), which includes both the dissociation constant of carbonic acid and the hydration constant (Eq. 2.9). Since $K_{hyd} \ll 1$, K_1 approximately equals the product of $K_a \times K_{hyd}$. During the last 20–30 years, intensive efforts have been directed at

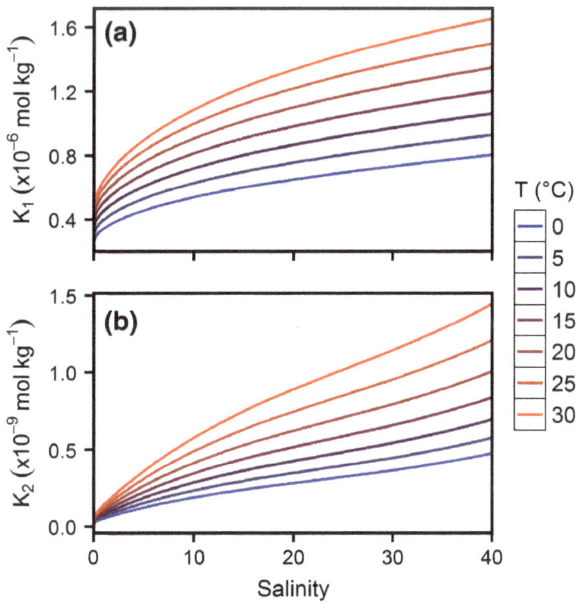

Fig. 2.3 Dissociation constants of the marine CO_2 system: **a** The first dissociation constants K_1 refers to CO_2^* (CO_2 + H_2CO_3) and **b** the second dissociation constant, K_2, to HCO_3^-

establishing equations that describe K_1 as a function of temperature, salinity, and pressure (an overview is given in Pierrot et al. 2006), with most focusing on the salinity range encountered in ocean waters. The only equation that also refers to brackish water and is consistent with the well-established K_1 for freshwater was published by Millero et al. (2006) and refined in Millero et al. (2010). The K_1 values obtained from this equation are plotted in Fig. 2.3a as a function of salinity (0–40) for temperatures between 0 and 20 °C. Within this range of conditions, the dissociation constant K_1 varies by almost one order of magnitude and, expressed as the negative decadal logarithm, is in the range of 6.6 > p K_1 > 5.8. Dissociation is enhanced at high temperatures and increases with increasing salinity. The salinity dependency is especially pronounced at salinities below 10, which are characteristic for the Baltic Proper and its adjacent gulfs.

During the second dissociation step, hydrogen carbonate ions transfer another proton to water molecules (Eq. 2.10):

$$HCO_3^- \Leftrightarrow CO_3^{2-} + H^+ \tag{2.10}$$

The corresponding equilibrium equation is shown in Eq. 2.11:

$$\frac{[CO_3^{2-}] \cdot [H^+]}{[HCO_3^-]} = K_2 \tag{2.11}$$

Qualitatively, the temperature and salinity dependencies of K_2 (Millero 2010) are similar to those of K_1. However, the range of K_2 is clearly larger, covering more than one order of magnitude (10.6 > pK_2 > 8.8) with respect to the limiting values used in Fig. 2.3b (**S** = 0–40, **T** = 0–30 °C).

Finally, the solubility of solid calcium carbonate has an impact on the marine CO_2 system. The formation of calcium carbonate shells by various marine organisms is a widespread phenomenon in ocean waters. However, in the Baltic Sea the abundance of calcifying plankton such as coccolithophores is significant only in the transition area between the Baltic Sea and the North Sea (Tyrrell et al. 2008). Biogenic calcium carbonate occurs in two different crystalline phases, either as calcite or aragonite. The solubility is characterized by the solubility product, K_{sp} (Eq. 2.12):

$$[Ca^{2+}] \cdot [CO_3^{2-}] = K_{sp} \tag{2.12}$$

Box 2.2: Derivation of the first dissociation constant of the marine CO_2 system, K_1

The equilibrium condition for the dissociation of the first hydrogen ion of carbonic acid constitutes the basis for the first dissociation constant, K_1, of the marine CO_2 system:

$$\frac{[H^+] \cdot [HCO_3^-]}{[H_2CO_3]} = K_a \tag{B2.1}$$

It is related to the hydration equilibrium:

$$\frac{[H_2CO_3]}{[CO_2]} = K_{hyd} \tag{B2.2}$$

Taking into account the definition of CO_2^*:

$$[CO_2^*] = [CO_2] + [H_2CO_3] \tag{B2.3}$$

and combining with B2.2, CO_2^* can be expressed by:

$$[CO_2^*] = \frac{[H_2CO_3]}{K_{hyd}} + [H_2CO_3] \tag{B2.4}$$

or:

$$[H_2CO_3] = [CO_2^*] \cdot \frac{K_{hyd}}{1 + K_{hyd}} \tag{B2.5}$$

Introducing B2.5 into B2.1, yields

$$\frac{[H^+] \cdot [HCO_3^-]}{[CO_2^*]} = K_a \cdot \frac{K_{hyd}}{1 + K_{hyd}} \tag{B2.6}$$

Hence, the conversion of CO_2^* to H^+ and HCO_3^- is characterized by an equilibrium constant, K_1, that is a composite of K_a and K_{hyd}:

$$\frac{[H^+] \cdot [HCO_3^-]}{[CO_2^*]} = K_1 \tag{B2.7}$$

$$K_1 = K_a \cdot \frac{K_{hyd}}{1 + K_{hyd}} \tag{B2.8}$$

In the case of calcite, K_{sp} is almost independent of temperature but increases strongly with increasing salinity, as shown in Fig. 2.4a. The change of K_{sp} with

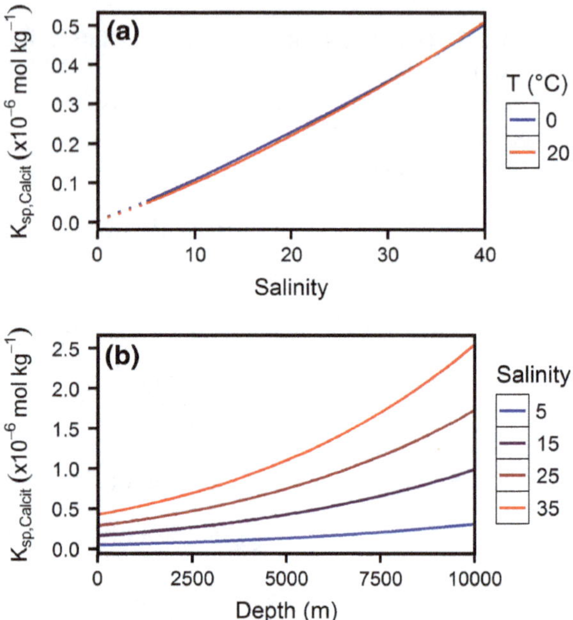

Fig. 2.4 The solubility product of calcite **a** as a function of salinity at different temperatures in surface water and **b** as a function of water depth at different salinities and T = 5 °C

pressure, expressed in terms of water depth in Fig. 2.4b, is even more pronounced. The calcite solubility at a depth of 10,000 m (T = 5 °C; S = 35) is 750-fold higher than in river water (T = 20 °C; S = 0). The solubility of aragonite is larger and in dependence on the physical conditions, the K_{sp} of aragonite exceeds that of calcite by a factor of 1.5–1.6.

Chemical equilibria between ionic CO_2 species are established almost spontaneously and also the hydration of CO_2, although slower, still occurs on a time scale of only seconds. By contrast, the dissolution of calcite or aragonite is a significantly slower process (Milliman 1975), resulting in the delayed establishment of the respective solubility equilibria. The degree of saturation (Ω) is defined as the ratio between the actual product of the concentration of Ca^{2+} and $CO_3{}^{2-}$ ions and the solubility product, as shown in Eq. 2.13:

$$\Omega = \frac{[Ca^{2+}] \cdot [CO_3^{2-}]}{K_{sp}} \tag{2.13}$$

Although ocean surface waters are generally oversaturated with calcite (Ω > 4), chemical precipitation of $CaCO_3$ does not occur because it is kinetically inhibited by magnesium ions (Berner 1975). Yet, calcifying organisms possess biochemical "tools" which enable them to overcome this inhibition and are even able to build $CaCO_3$ shells in waters that are undersaturated with $CaCO_3$. Considerable $CaCO_3$ oversaturation also characterizes river water and plays an important role in the input of alkalinity into the Baltic Sea (see Sect. 2.3.3).

2.3 Measurable Variables of the Marine CO_2 System

2.3.1 CO_2 Equilibrium Fugacity and Partial Pressure

Concentrations of CO_2^*, shown as a function of temperature and salinity in Fig. 2.2, refer to the equilibrium between the surface water and the atmospheric CO_2 fugacity (fCO_2). However, this equilibrium generally does not exist in nature because the gas exchange that is a prerequisite for its establishment always lags behind continuous changes in the surface-water equilibrium fCO_2 caused by changes in either temperature or the CO_2* concentration. In case of a temperature increase or decrease, both the solubility constant and, due to the dissociation equilibria, CO_2*, change. Hence, the equilibrium fCO_2 increases or decreases according to Eq. 2.14:

$$fCO_2 = \frac{[CO_2^*]}{K_0} \tag{2.14}$$

Similarly, any change in [CO_2*] that may occur as a result of biomass production or mineralization will affect the equilibrium fCO_2. The equilibrium fCO_2 is a fundamental property of a water mass and independent of the existence of a gas phase in contact with this water mass. For reasons of practicability, the equilibrium fCO_2 of seawater is commonly denoted simply as "fCO_2", in contrast to the atmospheric "fCO_2^{atm}". In the characterization of the marine CO_2 system, fCO_2 is an important variable because it is directly related to the concentration of CO_2* (Eq. 2.4), which yields HCO_3^- and CO_3^{2-} by the dissociation of carbonic acid. However, pCO_2 that at atmospheric pressure and composition deviates only very slightly from the fCO_2 is typically used to describe the status of the marine CO_2 system. Nonetheless, for further calculations concerning the CO_2 system, the use of software that converts pCO_2 to fCO_2 is recommended.

2.3.2 Total CO_2 and pH

C_T is the sum of the chemical species generated by the dissolution and dissociation of CO_2 in water and it is sometimes referred to as "dissolved inorganic carbon" (DIC) (Eq. 2.15):

$$C_T = [CO_2^*] + [HCO_3^-] + [CO_3^{2-}] \tag{2.15}$$

Typical C_T concentrations in pure water and surface seawater from the central Baltic and the ocean are shown in Fig. 2.5b. The bars in the figure represent the C_T at 10 °C and at equilibrium with an atmospheric pCO_2 of 400 µatm. The C_T of pure water (~25 µmol kg^{-1}) is 70- to 90-fold lower than that of seawater and ~90% is contributed by CO_2*; hence, it mainly consists of dissolved molecular CO_2.

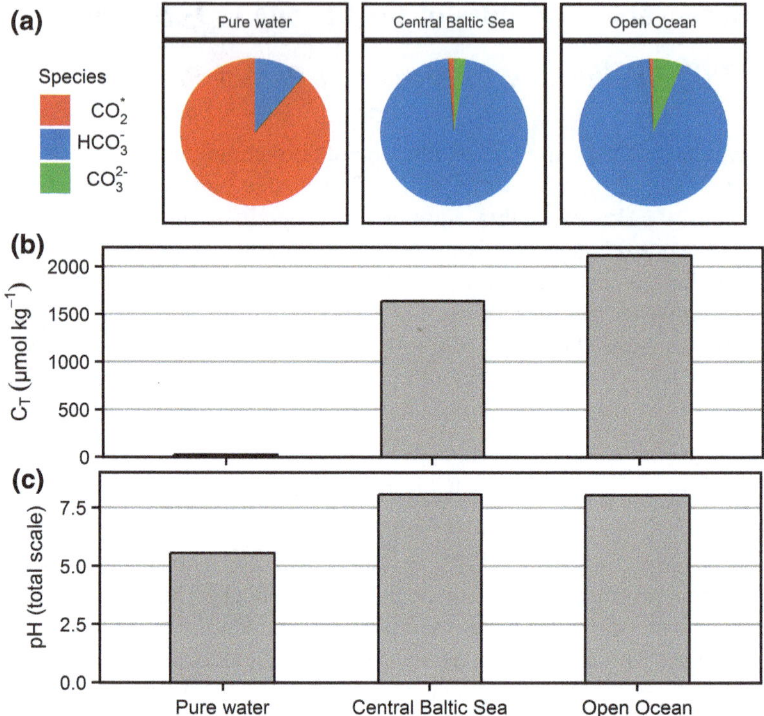

Fig. 2.5 Comparison between pure water, Baltic Sea water and ocean water concerning: **a** relative distribution of different CO_2 species, **b** the total amount of CO_2 (C_T), and **c** the pH (Computed for T = 10 °C, pCO_2 = 400 µatm; pure water: S = 0 and A_T = 0 µmol kg^{-1}, Baltic Sea: S = 7 and A_T = 1670 µmol kg^{-1}, ocean: S = 35 and A_T = 2300 µmol kg^{-1})

By contrast, in surface seawater, the CO_2* fraction accounts for only ~1% of C_T (red sections in Fig. 2.5a) and it is the C_T contribution of HCO_3^- (>90%; blue sections in Fig. 2.5a) that prevails. In pure water, the contribution of HCO_3^- is much lower (10%) and almost no (10^{-4}%) CO_3^{2-} ions are produced (green sections in Fig. 2.5a) during the dissolution of CO_2. Seawater is also low in CO_3^{2-}, but the contributions of the latter to C_T in the Baltic Sea and ocean water (3 and 6%, respectively) are significant and play an important biogeochemical role. The very large differences in the levels and composition of C_T between freshwater and seawater can be attributed to the alkalinity (A_T), which is another central variable in the marine CO_2 system (Sect. 2.3.3).

The concentrations of the different CO_2 species are coupled by the hydrogen ions generated during the stepwise dissociation of carbonic acid. The hydrogen ions control the two dissociation equilibria and, hence, the respective equations (Eqs. 2.8 and 2.11) include the hydrogen ion concentration and a reliable estimate of pH—the negative logarithm of the H^+ concentration—is an important piece of information to characterize the marine CO_2 system. However, the use of concentration units to describe dissociation equilibria conflicts with the traditional method for the determination of pH in seawater, which is based on potentiometric measurements performed with glass electrodes and on calibrations with pH buffers dissolved in pure water (NBS buffer). While this has been the standard procedure for many decades and is still common practice in many fields of environmental research, it has several shortcomings with respect to pH determinations in seawater because the obtained pH refers to the so-called NBS scale and reflects neither the hydrogen ion activity (Box 2.1) nor the hydrogen ion concentration (Dickson 1984). To overcome this conceptual problem, pH buffers prepared in artificial seawater were introduced. This leads to pH scales in which the standard reference state for the hydrogen ion activity no longer refers to the thermodynamic properties in pure water but to a solution of hydrogen ions in artificial standard seawater of defined salinity (ionic media scale). Hence, any interactions between hydrogen ions and seawater ions are no longer accounted for by the activity coefficient but are characteristics of the solute-solvent interactions. This convention allows the activity coefficient of the hydrogen ions to be set to unity and leads to a pH scale based on concentration units.

Since sulfate is a major seawater component, it should be included in the composition of the artificial seawater used as the solvent for buffer solutions. However, this implies that the interaction of hydrogen ions with the solvent is not confined to the protonation of water molecules, e.g., by the formation of hydronium ions (H_3O^+), but also includes the formation of hydrogen sulfate (HSO_4^-). The sum of the two is called the "total hydrogen ion" concentration (Eq. 2.16):

$$[H_T^+] = [H_F^+] + [HSO_4^-] \tag{2.16}$$

where H_F^+ represents the protonated water molecules referred to as "free hydrogen ions." If the artificial seawater contains fluoride, the formation of HF must also be taken into account and yields the so called seawater pH scale: $[H_{SWS}^+] = [H_F^+] + [HSO_4^-] + [HF]$, which, however, has been largely replaced by the H_T^+ concentration scale.

Using the HSO_4^- dissociation constant, K_S, and assuming a constant sulfate concentration, $[SO_4^{2-}]$, that depends only on salinity, H_T^+ can be expressed as shown in Eq. 2.17:

$$[H_T^+] = [H_F^+] \cdot (1 + \frac{[SO_4^{2-}]}{K_S}) \tag{2.17}$$

In ocean water (S = 35), by definition $[H_{SWS}^+] > [H_T^+] > [H_F^+]$, which is reflected in a pH_T that is ~0.09 units lower than pH_F, and pH_{SWS}, which in turn is about 0.01 units below the pH_T at 20 °C. These differences decrease with decreasing salinity. An exact thermodynamic determination of the differences between the NBS scale and the concentration scales at different salinities is not possible, but a difference of 0.13 pH units at a salinity of 35 is a reasonable estimate. In the following sections, we consistently refer to the pH on the total scale (pH_T) and use equilibrium constants that are based on it.

The pH_T of surface seawater is ~2 units higher than that of pure water at equilibrium with atmospheric CO_2 (Fig. 2.5c). Furthermore, the seawater CO_2 system is characterized by H^+ ion concentrations of ~10^{-8} mol kg^{-1}, which is 3–5 orders of magnitude lower than the concentrations of the other dissociation products, HCO_3^- and CO_3^{2-}. Hence, the dissociation equilibria are very sensitive to $[H^+]$ perturbations by the direct input of an acid or by the addition/removal of CO_2 that results in carbonic acid generation. The re-equilibration of the CO_2 system consequently requires the removal of the bulk of the added H^+ ions, mainly by the protonation of CO_3^{2-} ions.

Measurements of C_T and pH may be used to calculate the concentrations of the different CO_2 species in seawater, provided that the dissociations constants K_1 and K_2 (Eqs. 2.8 and 2.11) as a function of temperature and salinity are known. Similarly, the pCO_2, which is related to CO_2^* by the solubility constant K_0, in combination with pH or C_T allows the full determination of the marine CO_2 system.

2.3.3 Alkalinity

Alkalinity is another central quantity for the characterization of the marine CO_2 system. It merits a separate section because: (i) it not only impacts the marine CO_2 system, but also refers explicitly to all acid-base constituents in seawater and (ii) it constitutes the link between biogeochemical processes on land and those in the sea. In the early stages of research on the marine CO_2 system, alkalinity was also termed the "acid binding capacity." This more illustrative term reflects the precise definition that considers alkalinity as the excess of proton acceptors over proton donors. In this context, proton acceptors are the anions of weak acids ($K_a < 10^{-4.5}$ mol kg^{-1}) whereas proton donors are strong acids ($K_a > 10^{-4.5}$ mol kg^{-1}) (Dickson 1981). For ocean surface waters, total alkalinity (A_T) is sufficiently represented by Eq. 2.18:

$$A_T = [HCO_3^-] + 2[CO_3^{2-}] + [B(OH)_4^-] + [OH^-] - [H_T^+] \tag{2.18}$$

where, in addition to HCO_3^- and CO_3^{2-} (carbonate alkalinity), borate and hydroxide ions are included as proton acceptors and the total hydrogen ion concentration, H_T^+, represents strong acids, such as protonated water molecules and HSO_4^- ions (Eq. 2.18). Alkalinity may be determined by different methods of titration using a strong acid. In the direct potentiometric titration with HCl, the excess of proton acceptors is compensated for by the amount of protons added during the titration and an analysis of the titration curve yields the equivalence point, indicated by an inflection point.

In the characterization of the CO_2 system using the measured A_T, non-CO_2 A_T contributions are expressed as a function of pH. In a simplified representation of A_T according to Eq. 2.18, this yields (Eq. 2.19):

$$A_T = [HCO_3^-] + 2[CO_3^{2-}] + B_T \cdot \frac{K_B}{[H_T^+] + K_B} + \frac{K_W}{[H_T^+]} - [H_T^+] \tag{2.19}$$

Since the boric acid dissociation constant, K_B, and the ion product of water, K_W, are thermodynamic constants that are known as functions of temperature and salinity, and since the total boric acid (H_3BO_3 + $B(OH)_4^-$) concentration, B_T, in seawater is a function of salinity, A_T is exclusively described by variables that characterize the marine CO_2 system (Eq. 2.19). Therefore, A_T together with any other variable (either pCO_2, pH, or C_T) and the carbonic acid (CO_2) dissociation constants provides a full characterization of the marine CO_2 system. In semi-enclosed marine systems such as the Baltic Sea and the Black Sea, where the development of anoxic conditions in deeper water layers is a common feature, the accumulation of phosphate and the formation of S^{2-} (at equilibrium with H_2S, HS^-) and ammonia also contribute significantly to the pool of proton acceptors and must be accounted for in Eq. 2.19. In this case, the acid-base components are independent of salinity and their total concentrations must be obtained from measurements. This also applies to any other additional alkalinity contributions, such as organic proton acceptors (Kulinski et al. 2014).

Another important aspect of the occurrence of alkalinity is the question for the processes that produce the excess of proton acceptors in seawater. Or in other words: what are the sources and sinks of alkalinity? The dissolution of gaseous CO_2 in pure water or in an acid/base-free seawater medium and the subsequent dissociation of carbonic acid cannot generate alkalinity because charge balance requires the production of equal amounts of proton acceptors and donors (Eq. 2.19):

$$\Delta[HCO_3^-] + 2\Delta[CO_3^{2-}] + \Delta[OH^-] = \Delta[H_T^+] \tag{2.20}$$

hence, $\Delta A_T = 0$. Instead, it is river water that transports an excess of HCO_3^- and CO_3^{2-}, generated by carbonate and silicate weathering, into the marine environment. Carbonate weathering is particularly important in the southern and south-eastern catchment regions of the Baltic Sea, where soils rich in limestone and organic matter (OM) favor the dissolution of solid calcium carbonate. CO_2 generated in soils from OM mineralization enhances the dissolution of carbonate minerals by driving the protonation of CO_3^{2-} ions to form HCO_3^-, thereby decreasing the degree of calcium carbonate saturation and triggering further dissolution of carbonates. The net reaction for the formation of alkalinity consisting mainly of HCO_3^- is given by Eq. 2.21:

$$CaCO_3 + CO_2 + H_2O \Rightarrow Ca^{2+} + 2HCO_3^- \tag{2.21}$$

Carbonate weathering is less important in the Baltic's Scandinavian catchment area, where siliceous primary rocks dominate the geological structures. Although silicates are also subjected to weathering by CO_2 and generate carbonate alkalinity, this process is much slower (Liu et al. 2011) than carbonate weathering. The complex chemical processes associated with CO_2-induced silicate weathering can be expressed by the following net reaction (Eq. 2.22):

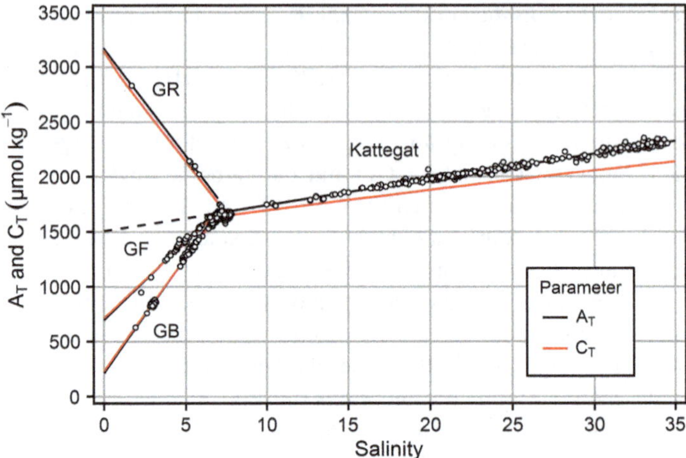

Fig. 2.6 Surface water alkalinity, A_T (*black*), and total CO_2, C_T (*red*), distributions as function of salinity in the Baltic Sea. The C_T levels are computed based on linear A_T-S-relations in the four sub-regions Kattegat, Gulf of Riga, Gulf of Finland and Gulf of Bothnia and assuming equilibrium with an atmospheric pCO_2 of 400 μatm and T = 10 °C (modified after Müller et al. 2016 with observations for the period 2008–2010)

$$CaSiO_3 + 2CO_2 + H_2O \Rightarrow Ca^{2+} + 2HCO_3^- + SiO_2 \tag{2.22}$$

The differences between the weathering processes in the Scandinavian and central European catchments are clearly indicated by the characteristics of the regional surface water salinity-alkalinity relationships (Fig. 2.6). The linear relationships in the three major Gulfs (Fig. 3.1) differ and extrapolating each one to zero salinity yields the flow-weighted mean A_T of river water entering the individual Gulfs. Due to the reduced weathering efficiency in the Scandinavian catchment, rivers entering the Gulf of Bothnia have the lowest A_T ($\sim$200 μmol kg^{-1}). An intermediate value is obtained for the Gulf of Finland ($\sim$700 μmol kg^{-1}), and a very high mean A_T ($\sim$3000 μmol kg^{-1}) for the Daugava River, which discharges into the Gulf of Riga. High A_T values were also directly observed in Odra and Vistula Rivers and can be attributed to intense carbonate weathering in central European catchments. The three regression lines intersect at a salinity of $\sim$7. This salinity is characteristic of large areas of the central Baltic Sea, which acts as a mixing chamber for water masses originating from the different gulfs, adjacent rivers, and inflowing North Sea water. Accordingly, also within the salinity interval ranging from 7 in the central Baltic Sea to $\sim$35 in the North Sea, a clear, linear, alkalinity-salinity relationship is obtained. Extrapolation of this line to a salinity of 0 yields an alkalinity of $\sim$1500 μmol kg^{-1}, which can be interpreted as the flow-weighted mean A_T of all rivers discharging into the Baltic Sea (Müller et al. 2016).

To explain the high A_T in continental European rivers by carbonate weathering requires a soil pCO_2 of several thousand μatm. Once the respective groundwater comes into contact with the atmosphere during its transport into the Baltic Sea, gas

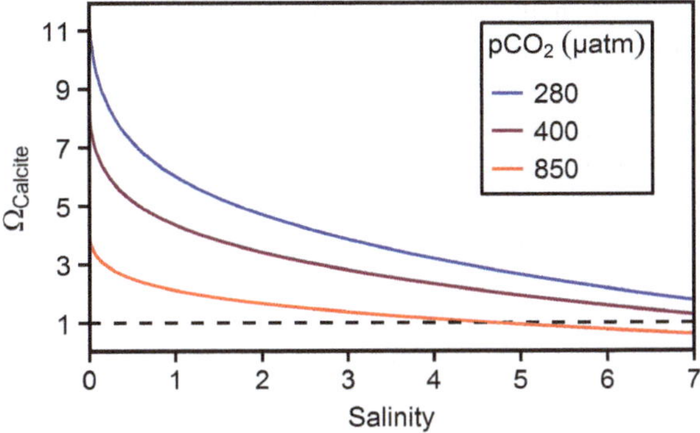

Fig. 2.7 Calcite saturation, $\Omega_{Calcite}$, as a function of salinity for the mixing of river water from the southern drainage basin (S = 0, A_T = 3000 µmol kg^{-1}) with Central Baltic Sea water (S = 7, A_T = 1650 µmol kg^{-1}). Calculations are performed for equilibrium with preindustrial (280 µatm), current (400 µatm) and worst case future (850 µatm) atmospheric pCO_2 levels and T = 5 °C

exchange occurs and the pCO_2 more or less adapts to the atmospheric pCO_2 level. During the release of CO_2, the CO_3^{2-} concentrations increase and the water becomes oversaturated with $CaCO_3$. Nonetheless, there is no precipitation of solid $CaCO_3$ because this process is inhibited, probably by the presence of magnesium ions (Berner 1975). However, the $CaCO_3$ oversaturation in river water decreases as the latter mixes with Baltic Sea water. This effect is illustrated in Fig. 2.7, which shows the degree of calcite saturation, $\Omega_{Calcite}$, as a function of salinity at a temperature of 5 °C for a river water A_T of 3000 µmol kg^{-1}. At the current atmospheric CO_2 content of ~400 ppm, the Ω for river water is ~8 but it decreases rapidly during the mixing of river water with Baltic Sea water and approaches 1, which implies equilibrium with solid calcite at a salinity of 7 in the central Baltic Sea.

The generation of alkalinity by weathering is intimately connected with an increase of the total CO_2 in groundwater because dissolved CO_3^{2-} ions are continually transformed to HCO_3^- by their reaction with CO_2 generated by the mineralization of soil OM. Hence, the dissolution of $CaCO_3$ and the subsequent formation of HCO_3^- continue until the groundwater has reached equilibrium with respect to both the dissociation of carbonic acid and the solubility of $CaCO_3$. The coupling of A_T and C_T at a given pCO_2 exists also in the Baltic Sea surface water, hence, at equilibrium with the atmospheric CO_2, A_T controls the C_T of seawater. As a result of this relationship, the plot of C_T versus S resembles that of A_T versus S (Fig. 2.6), although the relationship between A_T and C_T is not strictly linear.

2.3.4 Physico-Chemical Properties of the Master Variables

As total CO_2 (C_T) and total alkalinity (A_T) are conservative variables when related to mass units of seawater (e.g., µmol kg^{-1}) any changes in temperature or pressure will cause only a shift in the relative contributions of the chemical species that represent them (Eqs. 2.15 and 2.18) without affecting their magnitude. This conservative behavior is also reflected in the mass conservation that occurs during the mixing of two water masses (m_1, m_2) with different C_T or A_T values (c_1, c_2), as shown in Eqs. 2.23 and 2.24:

$$m_1 \cdot c_1 + m_2 \cdot c_2 = (m_1 + m_2) \cdot c_{mix} \tag{2.23}$$

or

$$c_{mix} = \frac{m_1 \cdot c_1 + m_2 \cdot c_2}{m_1 + m_2} \tag{2.24}$$

Due to their conservative properties, C_T and A_T are key variables in model simulations (e.g. Omstedt et al. 2014; Gustafsson 2012; Kuznetsov et al. 2011), where they are subjected to mixing and biogeochemical transformations and thus control changes in pCO_2 and in the hydrogen ion concentration, $[H^+]$. The latter variables are derived from C_T and A_T by the use of the equilibrium constants K_0, K_1, and K_2 (Eqs. 2.4, 2.8 and 2.11). Hence, pCO_2 and $[H^+]$ (expressed as pH) depend on temperature and pressure (depth) and are thus called "non-conservative." The non-conservative properties of $[H^+]$ are also reflected in its behavior during the

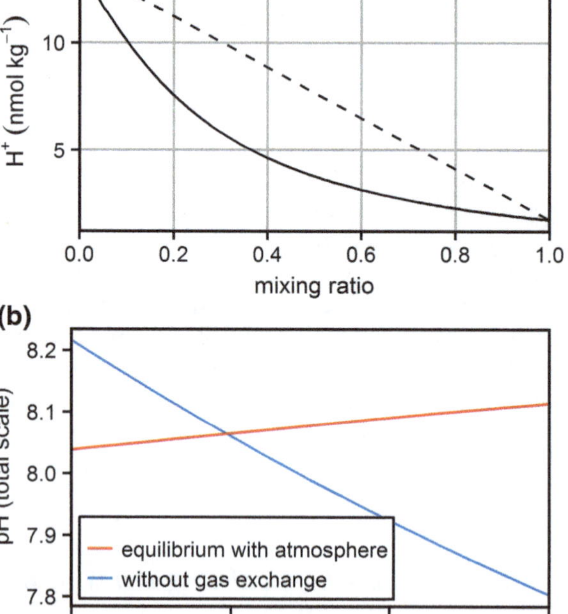

Fig. 2.8 Non-conservative behavior of the hydrogen ion concentration in seawater. **a** H^+ ion concentrations as a function of the mixing ratio of two water masses with different initial $[H^+]$, but the same A_T = 1670 μmol kg^{-1}, S = 7 and T = 10 °C (*solid line*). Hypothetical conservative mixing of $[H^+]$ is indicated by the *dashed line*. **b** Temperature dependency of pH during warming of the same water mass without gas exchange (C_T = const., 1640 μmol kg^{-1}) and in case that equilibrium with the ambient air is maintained (pCO_2 = const., 400 μatm)

mixing of different water masses. This is illustrated in Fig. 2.8a, which shows $[H^+]$ as a function of the mixing ratio of two water masses with the same temperature, salinity, and alkalinity, but different pH. The $[H^+]$ shows a distinct non-linear dependency on the mixing ratio and the differences with regard to a hypothetical conservative mixing line (Eq. 2.24) may be >30%.

The temperature dependency of non-conservative variables is especially important for the pCO_2 because along with biological activity it controls the seasonality of the surface water pCO_2, which in turn controls CO_2 gas exchange with the atmosphere. Furthermore, since in many cases pCO_2 cannot be measured at the in situ temperature, it is necessary to know the temperature coefficient to allow a correction. The temperature coefficient is commonly defined by the relative change of the pCO_2 per K, which is mathematically equivalent to $d(\ln pCO_2)/dT$. For oceanic systems, the experimentally derived coefficient of $0.0423\ K^{-1}$ (Takahashi et al. 1993), corresponding to a 4.23% change in pCO_2 per unit K, is a reasonable approximation. However, the values may be significantly lower (as low as $3\%\ K^{-1}$) at concurring low salinities and alkalinities, as observed in the Gulf of Bothnia (see Box 2.3 for details).

Box 2.3: Temperature dependency of pCO_2

The relative change in pCO_2 induced by a change in temperature, $d(\ln pCO_2)/dT$, was experimentally determined as $0.0423\ K^{-1}$ for largely stable and homogeneous oceanic conditions (Takahashi et al. 1993). This coefficient is used, for example, to estimate the reaction of pCO_2 upon the warming of ocean waters, or to correct the measured pCO_2 values to the in situ temperature. As for any other gas dissolved in water, the change in pCO_2 as a function of temperature is directly related to the change in the solubility constant K_0 (Eq. 2.14, Fig. 2.2). However, for CO_2 the temperature

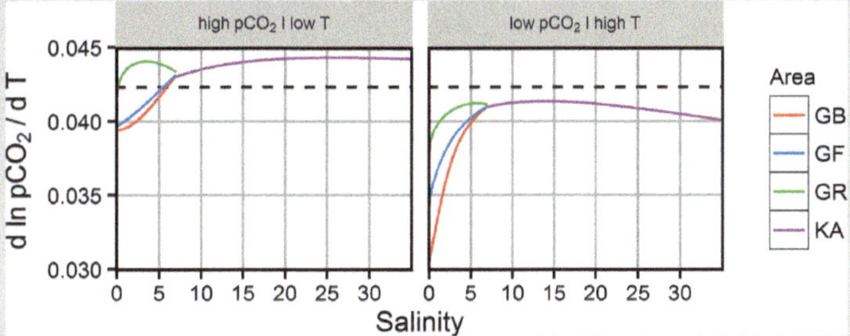

Fig. B3.1 The temperature (T) dependency of pCO_2 expressed as the relative change in pCO_2 per °K, $(d(\ln pCO_2)/dT)$. Values are given as a function of salinity for the different alkalinity and salinity regimes (compare with Fig. 2.6) encountered in the Gulf of Bothnia (GB), Finland (GF), and Riga (GR), as well as for the transition from the central Baltic Sea to the North Sea (the Kattegat, KA). The panels show typical winter (high pCO_2 = 500 μatm, low T = 5 °C) and summer (low pCO_2 = 300 μatm, high T = 20 °C) conditions. The dashed horizontal line indicates the experimentally derived value ($0.0423\ K^{-1}$) for oceanic conditions dependency of the partial pressure is also impacted by temperature-induced changes in the dissociation constants K_1 and K_2 that control the concentration of CO_2^*. Therefore, the coefficient depends on the compositions of the seawater, expressed by its salinity and alkalinity, and on the speciation of the CO_2 system components, controlled by the temperature and pCO_2. For the complex S, A_T, T, and pCO_2 conditions in the Baltic Sea, the coefficient d ($\ln pCO_2$)/dT can differ significantly from 0.0423 K^{-1}. Figure B3.1 presents the coefficient as a function of the salinity, calculated for the typical seawater composition of sub-regions of the Baltic Sea and for a combination of T and pCO_2 that roughly reflects winter and summer conditions. The deviations of the coefficient from the empirical oceanic value of 0.0423 K^{-1} are largest in the areas approaching the inner gulfs, for example, 0.036 K^{-1} (−15%) at S = 2 in the Gulf of Bothnia during summer. However, the salinity of the bulk of Baltic Sea surface water investigated in this book is between 5 and 10 and the corresponding deviation from oceanic value is <5% and thus in most cases negligible.

Likewise, the temperature dependency of pH is also a function of alkalinity and is influenced by salinity. In the absence of gas exchange with the atmosphere (C_T = constant), the pH of seawater decreases during warming. This is illustrated by the blue line in Fig. 2.8b, which shows the pH as a function of temperature at an alkalinity of 1600 μmol kg^{-1} and a salinity of 7, conditions typical for the central Baltic Sea. The pH decrease amounts to ~0.01 per K and corresponds to a relative increase in $[H^+]$ of ~2.3% per K. If the equilibrium with ambient air is maintained (pCO_2 = constant) during warming by spontaneous gas exchange, the loss of CO_2 reverses the temperature dependency of the pH and results in a slight increase of the pH.

To estimate qualitatively the effect of any biogeochemically or physically induced change in the marine CO_2 system, it is useful to consider the combined dissociation equilibria for carbonic acid. Dividing the equilibrium equation for the first dissociation step (Eq. 2.8) by that for the second dissociation step (Eq. 2.11), yields the relationship shown in Eq. 2.25:

$$\frac{[HCO_3^-]^2}{[CO_2^*]\cdot[CO_3^{2-}]} = \frac{K_1}{K_2} \tag{2.25}$$

This equation refers formally to the reaction described by Eq. 2.26:

$$CO_2 + CO_3^{2-} + H_2O = 2HCO_3^- \tag{2.26}$$

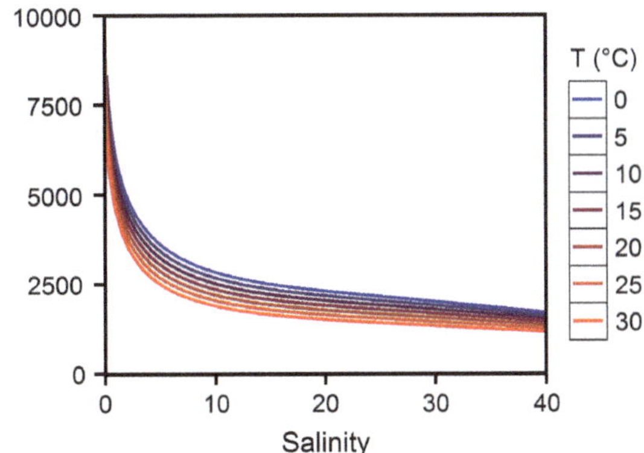

Fig. 2.9 Ratio of the dissociation constants K_1/K_2 as a function of salinity and temperature. The K_1/K_2 ratio formally corresponds to the equilibrium constant of the CO_2 buffer reaction (Eq. 2.25)

which is also known as "buffer reaction" in the context of the oceanic uptake of anthropogenic CO_2. The buffer equation does not account for the existence of hydrogen ions, which are involved in any reaction concerning the CO_2 system. However, the concentrations of H^+ are several orders of magnitude lower than those of the CO_2 species. Hence, any re-equilibration of the CO_2 system after the addition/removal of CO_2 or CO_3^{2-} occurs mainly by a shift in the relative distribution of the CO_2 species, which can be estimated from Eq. 2.25 or from the underlying reaction (Eq. 2.26). Thus, during the uptake of anthropogenic CO_2, most of the dissolved anthropogenic CO_2 is converted to HCO_3^- (buffered) by its reaction with CO_3^{2-}. Furthermore, the temperature and salinity dependency of K_1/K_2 allows an assessment of the corresponding changes in the relative distribution of the CO_2 species. Both increasing temperature and, in particular, increasing salinity lower the K_1/K_2 ratio (Fig. 2.9) and, according to Eqs. 2.25 and 2.26, increase the concentrations of CO_2^* and of CO_3^{2-} at constant C_T. The presence of the latter ions favors the formation of calcium carbonate shells and supports the high abundances of calcifying organisms, e.g., corals, in the high-temperature and high-salinity sub-tropic ocean.

The Revelle factor, which describes the relative change in pCO_2 in response to the relative change of total CO_2, is a further useful quantity to characterize the marine CO_2 system; it is defined in Eq. 2.27:

$$\frac{d(\ln pCO_2)}{d(\ln C_T)} = R \tag{2.27}$$

Assuming that the global ocean surface is, on average, approximately at equilibrium with the atmospheric CO_2, the Revelle factor can be used to assess the capacity of ocean surface water to take up anthropogenic CO_2, because R indicates how much CO_2 must be taken up by the water to achieve a new equilibrium with the atmospheric CO_2 after anthropogenic perturbation. Low R values imply that more anthropogenic CO_2 can be stored in the ocean at the expense of the fraction that stays in the atmosphere, and vice versa.

However, in many cases it is more informative to describe the relationship between C_T and pCO_2 by the differential quotient of their absolute values, R^* (Eq. 2.28):

$$\frac{dpCO_2}{dC_T} = R^* \tag{2.28}$$

In Chap. 5, changes in C_T are calculated from pCO_2 measurements to estimate net OM production. The sensitivity of this approach is represented by R^*, since it quantifies the signal in the measured pCO_2 that is triggered by the loss of C_T. R^* is related to the ratio $[CO_2^*]/[CO_3^{2-}]$ which in turn is a complex function of A_T, pCO_2, temperature, and salinity. The magnitude of the surface-water R^* is presented as a function of the pCO_2 in Fig. 2.10 for the main sub-regions of the Baltic Sea. The considered pCO_2 range of 100–600 µatm approximately reflects the seasonal pCO_2 variability encountered in the Baltic Proper and adjacent coastal areas, and

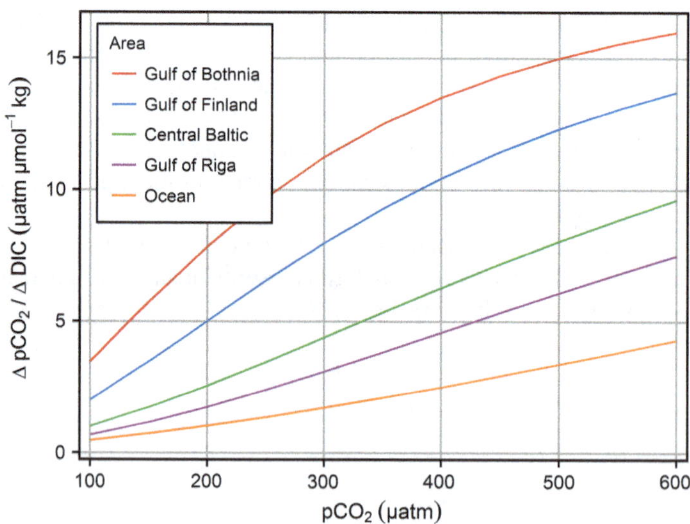

Fig. 2.10 Changes of pCO_2 upon changes in C_T ($R^* = \Delta pCO_2/\Delta C_T$) as a function of the pCO_2 level in different water masses: Gulf of Bothnia (S = 3, A_T = 800 µmol kg^{-1}), Gulf of Finland (S = 3, A_T = 1200 µmol kg^{-1}), Central Baltic (S = 7, A_T = 1670 µmol kg^{-1}), Gulf of Riga (S = 3, A_T = 2600 µmol kg^{-1}) and ocean water (S = 35, A_T = 2300 µmol kg^{-1}) at T = 10 °C. Highest R^* values result from low CO_3^{2-} concentrations, e.g., in water with low A_T and high pCO_2 levels

corresponds to an increase of R^* by a factor of 6–7. Similarly, alkalinity has a strong effect on R^*, leading to highest and lowest R^* values in the Gulf of Bothnia and ocean water, respectively. The regional and seasonal variability of R^* is thus considerable and has implications for the sensitivity and uncertainties associated with our approach to estimating changes in total CO_2 from pCO_2 measurements.

2.4 CO_2 Air-Sea Gas Exchange

In addition to its role in the characterization of the marine CO_2 system, pCO_2 is a major control of the CO_2 gas exchange between the sea surface and the atmosphere. The exchange at the sea surface of CO_2, O_2, N_2, or any other gas with a similarly low solubility is thermodynamically driven by the concentration gradient across the laminar layer at the air-sea interface. This layer, which is also referred to as the film layer or the viscous layer, is not influenced by turbulence and any transport of dissolved substances occurs by molecular diffusion. The thickness of the laminar layer depends on the physical state of the sea surface (wind speed) and is typically several tens of μm. Using a simple film model (Liss and Merlivat 1986), the flux, F, of any gas can be expressed as shown in Eqs. 2.29 and 2.30:

$$F = k \cdot (c^{sea} - c^{sat}) \tag{2.29}$$

or

$$F = k \cdot \Delta c \tag{2.30}$$

where c^{sea} is the concentrations of the gas in bulk seawater and represents the concentration at the lower boundary of the laminar layer, and c^{sat} is the saturation concentration of the gas presumed to be present at the upper boundary of the laminar layer. This interpretation of the flux equation is based on the assumption that the exchange of gases with low solubilities is entirely controlled by transfer across the laminar layer at the water surface (water-side control). By contrast, the exchange of gases with high solubilities, as is the case for many volatile organic substances and water vapor itself, is controlled by the concentration gradient in the gas phase. If the exchanged gas is CO_2, c^{sea} and c^{sat} refer to CO_2^*, which represents the gaseous fraction of the dissolved CO_2 species. However, since CO_2^* cannot be measured directly and is only accessible by measurements of pCO_2 (Eq. 2.14), it is convenient to express Eq. 2.30 as Eq. 2.31:

$$F = k \cdot K_o \cdot (pCO_2 - pCO_2^{atm}) \tag{2.31}$$

or

$$F = k \cdot K_o \cdot \Delta pCO_2 \tag{2.32}$$

where ΔpCO_2 is the partial pressure difference between the surface water $pCO_2$and the atmospheric pCO_2. It is commonly used to characterize the CO_2 saturation of seawater with regard to atmospheric CO_2 although the fugacity difference (ΔfCO_2) is the correct thermodynamic representation.

The dynamic of the gas exchange is given by the transfer velocity k, which depends on the diffusion coefficient of the gas in seawater and on the thickness and stability of the laminar layer. The latter are influenced by microphysical processes, most of which are wind-controlled. Therefore, most of the existing bulk parameterizations of k are based on wind speed and many empirical equations have been suggested that describe k, at standard conditions of 20 °C and a salinity of 35, as a function of the wind speed, u (Wanninkhof 1992). The conversion of k to any other temperature and salinity requires the Schmidt number, Sc (Wanninkhof 1992), which is defined as the ratio between the kinematic viscosity and the diffusion coefficient. Although both are a function of temperature and salinity, variations of Sc for seawater are predominantly caused by temperature, because of the pronounced temperature dependency of the diffusion coefficient. By contrast, the effect of salinity on Sc is low and can generally be ignored. A reciprocal proportionality between the square root of Sc and k is assumed. Since the value of Sc at 20 °C and a salinity of 35 is 660, the transfer velocity k at any Sc calculated from the respective temperature and salinity is given by Eq. 2.33:

$$k = k_{660}(u) \cdot \left(\frac{660}{Sc(T,S)}\right)^{0.5} \tag{2.33}$$

The exponent 0.5 is not a physical constant but a widely used experimental approximation. For the CO_2 flux calculations in the following sections we use the quadratic equation for $k_{660}(u)$ shown in Eq. 2.34 and suggested by Wanninkhof et al. (2009):

$$k_{660} = 0.24u^2 \tag{2.34}$$

The coefficient of 0.24 applies to wind speed in m s^{-1} and yields k_{660} in cm h^{-1}.

Equations 2.31–2.34 are used to determine the CO_2 flux for a given partial pressure difference and to estimate the balance of the air-sea CO_2 gas exchange on the basis of pCO_2 time series data. It is based on the film model that accounts for the molecular diffusion of CO_2 through the laminar layer. However, a gradient of CO_2 across the laminar layer also implies gradients for HCO_3^- and CO_3^{2-} if equilibria within the CO_2 system are established. Thus, an additional net C_T flux occurs by the diffusion of HCO_3^- and CO_3^{2-} that can be accounted for by introducing a chemical enhancement factor into Eq. 2.33. However, the establishment of the CO_2 hydration equilibrium (Eq. 2.2) at the interface between the film layer and the atmosphere is relatively slow and at higher wind speeds lags behind the continuous renewal of the surface layer by turbulence. This hampers the formation of additional HCO_3^- and CO_3^{2-} gradients across the film layer such that a significant increase (>10%) of the air-sea CO_2 flux by chemical enhancement can only be expected at wind speeds

<4 m s^{-1} (Kuss and Schneider 2004). However, these estimates are associated with considerable uncertainty; thus, chemical enhancement of the CO_2 gas exchange is usually neglected in view of other uncertainties in the CO_2 flux calculations that are mainly due to the determination of k_{660}.

A further important characteristic of the air-sea gas exchange of any dissolved gas is the equilibration time, which characterizes the decrease in the concentration difference (Eq. 2.29) or the partial pressure difference (Eq. 2.32) across the film layer during gas exchange until a final equilibration with the ambient air is reached. This requires integration over time of the flux equation (Eq. 2.29), which can easily be performed for gases such as O_2 or N_2 that do not react with water (Box 2.4). Assuming that the atmospheric concentration of the considered gas is not affected by the gas exchange (c^{sat} = constant, see Eq. 2.29), the integration from t = 0 to any time t yields Eq. 2.35:

$$\Delta c(t) = \Delta c(t = 0) \cdot \exp - \frac{k}{z_{mix}} \cdot t \quad (2.35)$$

where z_{mix} represents the depth of the mixed layer within which the gas gain or loss caused by gas exchange is homogenously distributed. According to the film model, the gas concentration in the mixed layer corresponds to the concentration at the lower boundary of the laminar layer.

According to Eq. 2.35, a perfect equilibrium between the water and the atmosphere ($\Delta c = 0$) is only achieved after an infinite time. This is a consequence of the mathematical idealization of the involved processes. To nonetheless quantify the dynamic of the equilibration process, the ratio z_{mix}/k, which has the unit "time," is used to characterize the equilibration process and is called the equilibration time (τ), described in Eqs. 2.36 and 2.37:

$$\tau = \frac{z_{mix}}{k} \quad (2.36)$$

$$\Delta c(t) = \Delta c(t = 0) \cdot \exp -t/\tau \quad (2.37)$$

Hence, if the elapsed time after the start of the equilibration process is equal to τ, then $\Delta c(t)/\Delta c(t = 0)$ is equal to $e^{-1} = 0.36$. In other words, τ is the time after which 36% of the original $\Delta ct = 0$ still exists or, conversely, 64% of the original disequilibrium is compensated by gas exchange. Therefore, τ is also called the e-fold equilibration time. For O_2 gas exchange at a mean wind speed of 7 m s^{-1}, a mixed-layer depth of 30 m, and a temperature of 10 °C, τ is equal to 7 days.

Box 2.4: Integration of the flux equation

Here we consider the air-sea exchange of a gas that does not react with water (O_2, N_2, etc.). The basic equation used to calculate the flux (Eq. 2.29) is:

$$F = k \cdot (c^{sea} - c^{sat}) \tag{B4.1}$$

where the flux, F, is defined by the number of moles, dn, that pass the air-sea interface per time, dt, and area, A:

$$F = \frac{dn}{dt \cdot A} \tag{B4.2}$$

Hence, B4.1 can be written as:

$$\frac{dn}{dt \cdot A} = k \cdot (c^{sea} - c^{sat}) \tag{B4.3}$$

dividing by the depth of the mixed layer, z_{mix}, gives:

$$\frac{dn}{dt \cdot A \cdot z_{mix}} = k/z_{mix} \cdot (c^{sea} - c^{sat}) \tag{B4.4}$$

where $A \times z_{mix}$ represents the water volume that is affected by the addition/removal of dn moles of the gas. Hence, the gas exchange can be expressed in terms of the concentration changes in bulk seawater, dc^{sea}:

$$-\frac{dc^{sea}}{dt} = k/z_{mix} \cdot (c^{sea} - c^{sat}) \tag{B4.5}$$

According to the definition of Δc (Eq. 2.29), the flux will have a positive sign if it is directed from the water into the atmosphere. Hence, the change in the concentration in the water, dc^{sea}/dt, will have a negative sign. For the integration of Eq. 2.30, it is assumed that no other processes other than gas exchange affect Δc and that the change in the atmospheric concentration caused by the gas flux is negligible (c^{sat} = constant). The latter assumption implies that dc = d(Δc):

$$-\frac{d(\Delta c)}{dt} = k/z_{mix} \cdot \Delta c \tag{B4.6}$$

which can also be expressed as:

$$d(\ln \Delta c) = -k/z_{mix} \cdot \mathrm{dt} \tag{B4.7}$$

The integration of which yields:

$$\ln \frac{\Delta c(t)}{\Delta c(t = 0)} = -k/z_{mix} \cdot t \tag{B4.8}$$

or

$$\Delta c(t) = \Delta c(t = 0) \cdot \exp -k/z_{mix} \cdot t \tag{B4.9}$$

The description of the equilibration process by gas exchange is more complicated for CO_2. The flux calculations are based on the diffusion of CO_2^*, but changes in the concentrations must refer to C_T since the CO_2^* flux affects the concentrations of all CO_2 species according to the CO_2 dissociation equilibria (Eqs. 2.8, 2.11, 2.26). To maintain the equilibrium between the CO_2 species in water, the major fraction of the CO_2 added by gas exchange in case of undersaturation must react with CO_3^{2-} to form HCO_3^-. Conversely, in case of CO_2 oversaturation, the major fraction of the CO_2 removed from the water is replaced by the conversion of HCO_3^- to CO_2 and CO_3^{2-} (Eq. 2.26). Therefore, only a certain fraction of the exchanged CO_2* contributes to the reduction of the disequilibrium that is reflected in the CO_2^* gradient across the laminar surface layer. This means that c^{sea} in the flux equation (Eq. 2.29) refers to CO_2^*, whereas the concentration change caused by the flux F must be described in terms of total CO_2, C_T, as described in Eqs. 2.38–2.41:

$$F = k \cdot ([CO_2^{*,sea}] - [CO_2^{*,sat}]) \tag{2.38}$$

or (see Box 2.4):

$$\frac{dC_T}{dt} = k/z_{mix} \cdot ([CO_2^{*,sea}] - [CO_2^{*,sat}]) \tag{2.39}$$

introducing the variable, f_D (dimensionless delay factor):

$$f_D = \frac{d[CO_2^{*,sea}]}{dC_T} \tag{2.40}$$

yields:

$$\frac{d[CO_2^{*,sea}]}{dt} = f_D \cdot k/z_{mix} \cdot ([CO_2^{*,sea}] - [CO_2^{*,sat}]) \tag{2.41}$$

The variable f_D is, by definition, equal to 1 for non-reactive gases and <1 in case of CO_2 dissolved in seawater. We will refer to it as the delay factor in the equilibration of the marine CO_2 system with atmospheric CO_2 since it reduces the term k/z_{mix}, previously defined as the reciprocal equilibration time for non-reactive gases. However, f_D is not a constant but rather a complex function of the state of the CO_2 system. As such, it changes during the equilibration process. This implies that there is no strict exponential decay of ΔpCO_2 during equilibration, nor exists a well-defined equilibration time as for non-reactive gases.

However, to illustrate the equilibration of the CO_2 system, the gas-exchange process has been numerically simulated based on Eq. 2.41 (Box 2.4). The calculations derive from the same scenario used to determine the equilibration time for O_2 (wind speed = 7 m s^{-1}, z_{mix} = 30 m, sea surface temperature = 10 °C, S = 7). Assuming an initial air-sea disequilibrium, ΔpCO_2, the temporal development of the equilibration process by CO_2 gas exchange was calculated for finite time intervals. Since the dynamics of the equilibration depend on the delay factor (Eqs. 2.40 and 2.41) and thus on the state of the marine CO_2 system, separate calculations were performed for the five different water masses with characteristic alkalinity and salinity regimes: the Gulfs of Bothnia, Finland and Riga, the central Baltic Sea, and ocean water. The curves in Fig. 2.11 show the decrease in ΔpCO_2 over time, expressed as a percentage of the initial value. The equilibration curve for O_2 (Eq. 2.35), which must be the same for the different regions, is also included in the figure. After passage of the e-fold equilibration time τ, 63% of the original disequilibrium is compensated by gas exchange (horizontal line). For the selected conditions and in the case of oxygen, τ is roughly equal to 7 days. Although CO_2 does not have a defined e-fold equilibration time, the 63% equilibrium level is also used for it. In all cases τ is much larger for CO_2 than for O_2, with differences varying by factors between 3 (Gulf of Bothnia) and 10 (ocean water). This implies that any changes in the CO_2 system due to OM production or mineralization are conserved for a longer time in surface water than is the corresponding O_2 signal. This finding underlines the advantage of using CO_2 instead of O_2 measurements for biogeochemical studies.

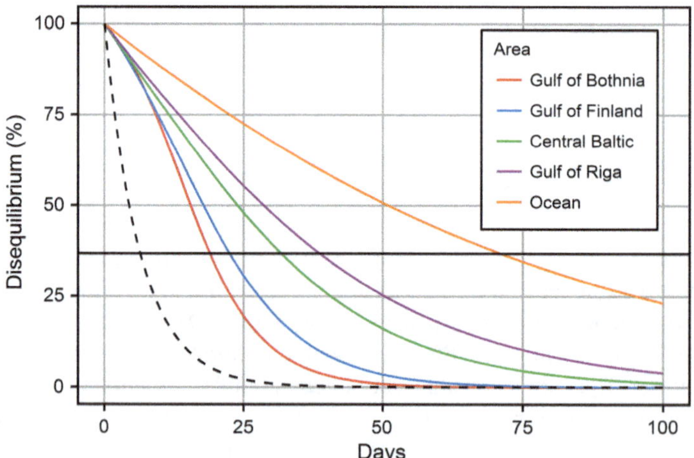

Fig. 2.11 Equilibration of O_2 (*black dashed line*) and of CO_2 (*colored lines*) with atmospheric O_2 and current CO_2. The equilibration of O_2 is characterized by an e-fold equilibration time after which 67% of the original horizontal disequilibrium is compensated for by gas exchange (*black horizontal line*). The equilibration of CO_2 is delayed and depends at a given pCO_2 mainly on the alkalinity. For the calculations we used the following parameters: Gulf of Bothnia (S = 3, A_T = 800 µmol kg^{-1}), Gulf of Finland (S = 3, A_T = 1200 µmol kg^{-1}), Central Baltic (S = 7, A_T = 1670 µmol kg^{-1}), Gulf of Riga (S = 3, A_T = 2600 µmol kg^{-1}) and ocean water (S = 35, A_T = 2300 µmol kg^{-1}) at T = 10 °C. The initial pCO_2 disequilibrium was in all sub-regions equivalent to a C_T drawdown of 50 µmol kg^{-1} from equilibrium with atmospheric pCO_2

In addition, the differences between the τ values for CO_2 in the different water masses are remarkable. Under the conditions defined in Fig. 2.11, the τ of ocean water is more than three times higher than that of water in the Gulf of Bothnia. This is due to the dependency of the delay factor f_D (Eq. 2.40) on the carbonate ion concentration, which is low in the Gulf of Bothnia because of the low alkalinity. Furthermore, the salinity affects the carbonate ion concentration (Fig. 2.9) because of greater dissociation at high salinities. Accordingly, τ is highest for ocean water although the highest alkalinity occurs in the Gulf of Riga. It must be noted that the equilibration curves in Fig. 2.11 demonstrate only the relative differences in τ for O_2 and CO_2 in water masses with differing alkalinity and salinity. To assess absolute τ values requires that other variables, such as the mixed-layer depth and wind speed (Eqs. 2.36 and 2.34), are taken into account. It must also be noted that the equilibration time is a theoretical quantity because under natural conditions disequilibria are incessantly generated by changing temperatures and biogeochemical processes. Nevertheless, it is a very useful quantity for understanding the response of the marine CO_2 system to the biogeochemical consumption or release of CO_2.

References

Berner RA (1975) The role of magnesium in the crystal growth of calcite and aragonite from sea water. Geochim Cosmochim Acta 39(4):489–504

Dickson AG (1984) pH scales and proton-transfer reactions in saline media such as sea water. Geochim Cosmochim Acta 48(11):2299–2308

Dickson AG (1981) An exact definition of total alkalinity and a procedure for the estimation of alkalinity and total inorganic carbon from titration data. Deep Sea Research Part A. Oceanographic Research Papers 28(6):609–623

Gustafsson E (2012) Modelling long-term development of hypoxic area and nutrient pools in the Baltic Proper. J Mar Syst 94:120–134

Kulinski K, Schneider B, Hammer K, Machulik U, Schulz-Bull D (2014) The influence of dissolved organic matter on the acid-base system of the Baltic Sea. J Mar Syst 132:106–115

Kuss J, Schneider B (2004) Chemical enhancement of the CO_2 gas exchange at a smooth seawater surface. Marine Chemistry, 91(1–4):165–174

Kuznetsov I, Neumann T, Schneider B, Yakushevm E (2011) Processes regulating pCO_2 in the surface waters of the central eastern Gotland Sea: a model study. Oceanologia 53:745–770

Lavigne H, Epitalon JM, Gattuso J-P (2011) Seacarb: seawater carbonate chemistry with R. http://cran.r-project.org/package=seacarb

Le Quéré C, Andrew RM, Canadell JG et al (2016) Global Carbon Budget 2016. Earth Syst Sci Data 8:605–649

Liss PS, Merlivat L (1986) Air-sea gas exchange: introduction and synthesis. In: Buat-Menard P (ed) The role of air-sea exchange in geochemical cycling. NATO ASI series, C 185, Reidel, pp 113–127

Liu Z, Dreybrodt W, Liu H (2011) Atmospheric CO_2 sink: silicate weathering or carbonate weathering? Appl Geochem 26(Supplement):S292–S294

Millero FJ, Graham TB, Huang F, Bustos-Serrano H, Pierrot D (2006) Dissociation constants of carbonic acid in seawater as a function of salinity and temperature. Marine Chemistry 100(1–2):80–94

Millero FJ (2010) Carbonate constants for estuarine waters. Mar Freshwater Res. 61:139–142

Milliman JD (1975) Dissolution of aragonite, Mg-calcite, and calcite in the North Atlantic Ocean. Geology 3:461–462. doi:10.1130/0091-7613(1975)

Müller JD, Schneider B, Rehder G (2016) Long-term alkalinity trends in the Baltic Sea and their implications for CO_2-induced acidification. Limnol Oceanogr 61:1984–2002. doi:10.1002/lno.10349 –open access–

Omstedt A, Humborg C, Pempkowiak J, Perttilä M, Rutgersson A, Schneider B, Smith B (2014) Biogeochemical control of the coupled CO_2–O_2 system of the Baltic Sea: a review of the results of Baltic-C. Ambio 43(1):49–59

Pierrot D, Lewis DE, Wallace DWR (2006) MS Excel program developed for CO_2 system calculations. ORNL/CDIAC-105a. Carbon Dioxide Information Analysis Center, Oak Ridge National Laboratory, U.S. Department of Energy, Oak Ridge, Tennessee. doi:10.3334/CDIAC/otg.CO2SYS_XLS_CDIAC105a

Schneider B, Gülzow W, Sadkowiak B, Rehder G (2014) High potential of VOS-based measurements in Baltic Sea surface waters for detecting sinks and sources of carbon dioxide and methane. J Mar Syst 140:13–25

Soli AL, Byrne RH (2002) CO_2 system hydration and dehydration kinetics and the equilibrium CO_2/H_2CO_3 ratio in aqueous NaCl solution. Mar Chem 78:65–73

Takahashi T, Olafsson J, Goddard JG, Chipman DW, Sutherland SC (1993) Seasonal variation of CO_2 and nutrient salts in the high latitude oceans: a comparative study. Glob Biogeochem Cycles 7:843–848

Tyrrell T, Schneider B, Charalampopoulou A, Riebesell U (2008) Coccolithophores and calcite saturation state in the Baltic and Black Seas. Biogeosciences 5:1–10

Wanninkhof R (1992) Relationship between wind speed and gas exchange over the ocean. J Geophys Res 97:7373–7382

Wanninkhof R, Asher WE, Ho DT, Sweeney C, McGillis WR (2009) Advances in quantifying air-sea gas exchange and environmental forcing. Annu Rev Mar Sci 1(1):213–244

Weiss RF (1974) Carbon dioxide in water and seawater: the solubility of a non-ideal gas. Mar Chem 2:203–215

Chapter 3
The Main Hydrographic Characteristics of the Baltic Sea

3.1 Water Budget and Estuarine Circulation

The water balance of the Baltic Sea can be assessed by subdividing the sea into sub-regions that represent the surface water and deep water of the Baltic Sea Proper (BP), the adjacent Gulfs of Bothnia (GB), Finland (GF) and Riga (GR), and the Kattegat (KA) displayed in Fig. 3.1.

Except for the Kattegat, a closed water budget is assigned to each compartment and includes river water input (Q_R), water exchange with adjacent compartments, precipitation (P), and evaporation (E). The latter two variables are not quantified explicitly, but by their difference (P–E). The Kattegat is not considered as a closed compartment; rather, it acts as a sink/source region for Baltic Sea water. As shown in Fig. 3.2, annual river-water input into the entire Baltic Sea is 450 km^3, whereof 25% (114 m^3) enters the Baltic Sea Proper directly and the remaining 75% via water exchange with the Gulfs. The annual contribution of the excess precipitation over evaporation to the freshwater input into the Baltic Sea Proper is small, $\sim$10 km^3. The high freshwater input triggers an estuarine circulation through which low-salinity surface water (Fig. 3.1 and 3.5), formed by the mixing of river water with high-salinity deep water, is exported to the Kattegat at a rate of 780 km^3 $year^{-1}$. This exceeds the inputs of both freshwater from rivers and the excess precipitation over evaporation by 280 km^3 $year^{-1}$. The difference corresponds to the input of Kattegat water to the deep Baltic Sea Proper, where it forms the sub-surface water layer and by mixing contributes to the salinity of the outflowing water.

The data for the water flows and inventories of the compartments are taken from Hjalmarsson et al. (2008) and from HELCOM (2006).

B. Schneider and J.D. Müller, *Biogeochemical Transformations in the Baltic Sea*,
Springer Oceanography, DOI 10.1007/978-3-319-61699-5_3

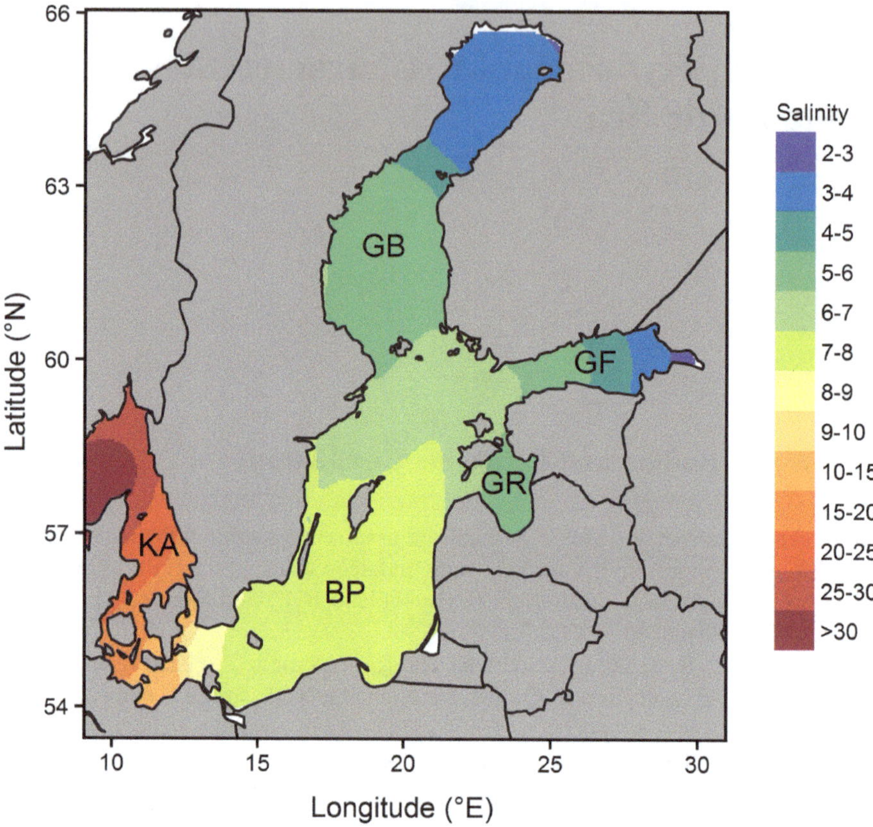

Fig. 3.1 The surface water salinity distribution and major sub-regions of the Baltic Sea: *BP* Baltic Proper, *GB* Gulf of Bothnia, *GF* Gulf of Finland, *GR* Gulf of Riga, *KA* Kattegat

3.2 Seasonality of the Stratification

The estuarine circulation implies a permanent salinity stratification that is most pronounced in the Baltic Sea Proper. The permanent halocline is currently located at a depth of 50–60 m in the eastern Gotland Sea and it has far-reaching importance in the cycling of biogeochemically active elements released by OM mineralization (see Sect. 3.3) in deeper water layers. In the production of OM by photosynthesis, the seasonal stratification of the surface water plays the more important role. It is mainly controlled by the heat balance of the surface water, freshwater inputs, and meteorological conditions. Figure 3.3 depicts the seasonality of the stratification by the vertical distribution of temperature and salinity at station BY15, in the central Gotland Sea (Fig. 4.4) in 2009, the year that was also used in the analysis of the surface water pCO_2, as described in Chap. 5.

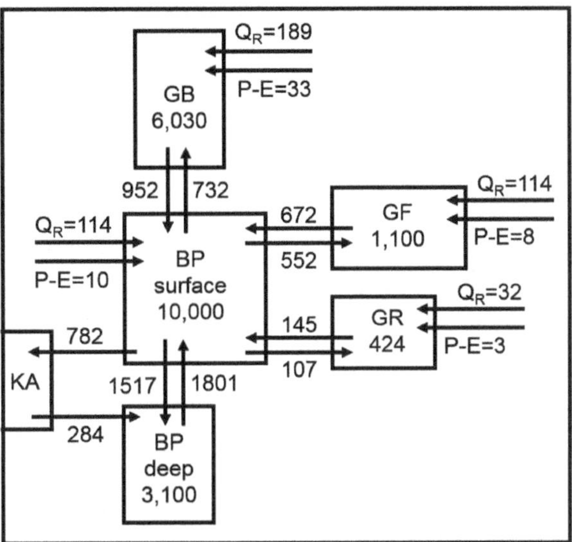

Fig. 3.2 The water budget of the Baltic Sea and its sub-regions: *BP* Baltic Proper, *GB* Gulf of Bothnia, *GF* Gulf of Finland, *GR* Gulf of Riga, *KA* Kattegat. Numbers in the compartments refer to the inventories (km^3), the *arrows* represent the water flows (km^3 $year^{-1}$) between the compartments, the input by river water (Q_R) and by precipitation corrected for evaporation (P–E)

In February, the surface water was fully mixed, down to the permanent halocline. Hence, temperature and salinity were homogenously distributed between the surface and a depth of about 60 m, where both temperature and salinity increased abruptly (Fig. 3.3). Until May, the surface water temperature increased by ~4 K; during this process a thermal mixed layer formed whose depth gradually decreased, reaching 20 m in May. The water between the thermocline and the permanent halocline is called the intermediate water or "winter water," because its physical properties are largely those acquired during the preceding winter. A further temperature increase of the surface layer by >10 K occured until July. Depending on the wind conditions, this may have generated different temperature distribution patterns in the upper surface layer. At low wind speeds (approximately <5 m s^{-1}) the temperature between the thermocline established in May and the air-sea interface increased gradually (Fig. 3.3, open red circles) and a true surface mixed layer was not detected at the given depth resolution of the data (5 m). This structure of the surface temperature distribution vanished at higher wind speeds and a distinct thermocline was re-established, which in 2009 was located at a depth of 15 m (Fig. 3.3, closed red circles), slightly shallower than in May. Finally, Fig. 3.2 shows the response of stratification to both the cooling of the surface water and the increasing wind speeds in autumn and winter. A thermocline was still present, but shifted to a depth of ~50 m and was thus close to the depth of the permanent halocline.

The salinity above the thermocline is also subjected to a seasonality, with minimum values generally reached in midsummer. Hence, the decreasing salinity supports the development of thermal stratification. However, the difference in

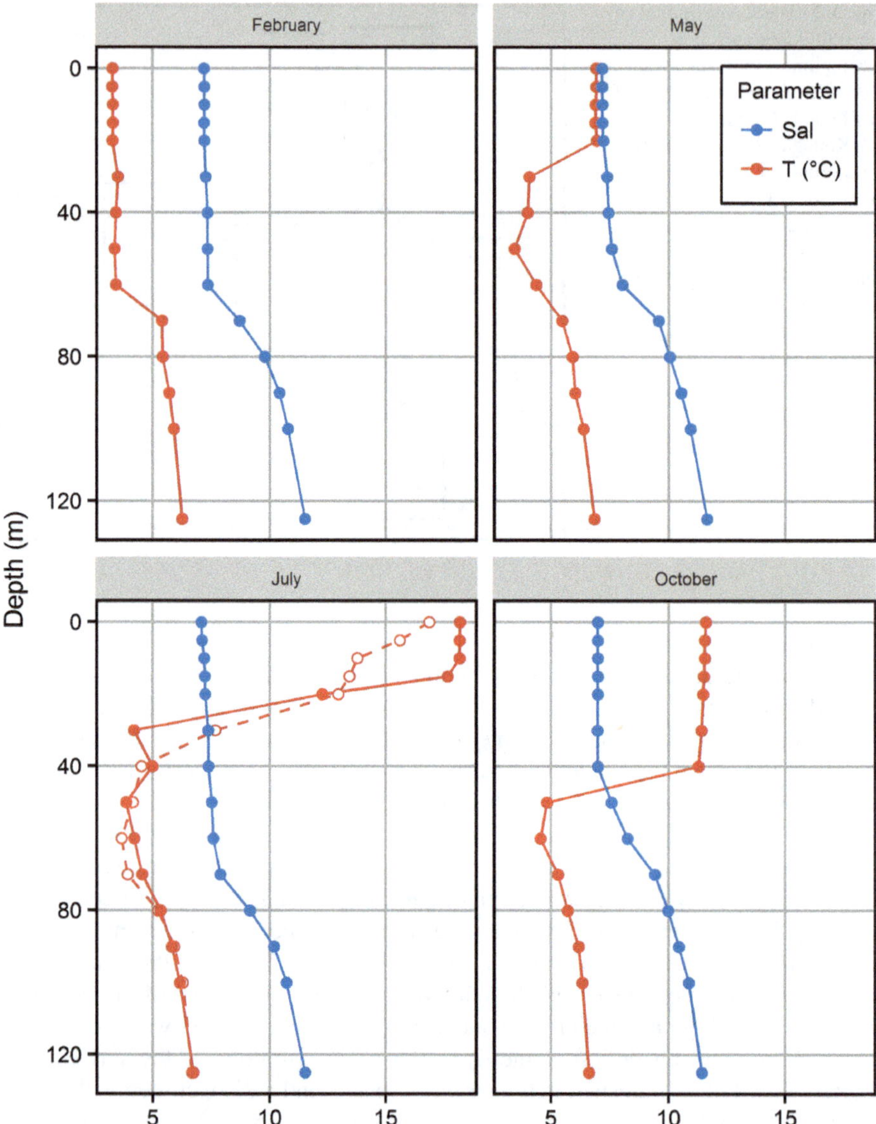

Fig. 3.3 The seasonality of the temperature and salinity stratification in the central Baltic Sea in 2009. Different temperature profiles were observed in July and refer to different wind conditions. During low wind speed the temperature (*open circles*) increased gradually towards the surface and a mixed surface layer was not formed

salinity between the surface mixed layer and the underlying intermediate water is only a few tenths of salinity units. Given the very large temperature differences at the thermocline (Fig. 3.3), temperature is the major control of surface-water density

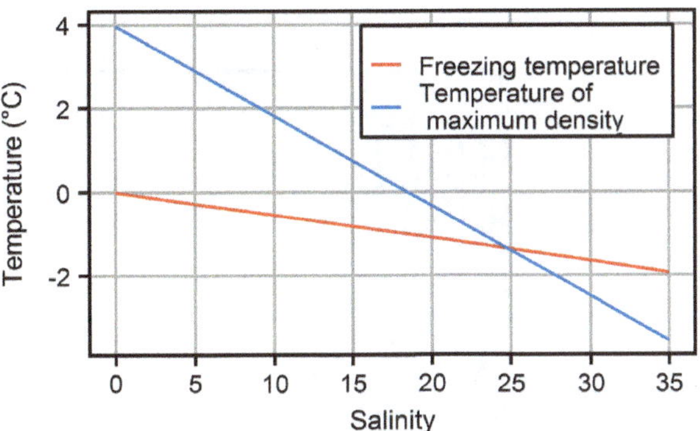

Fig. 3.4 Dependency of the temperature of maximum density and the freezing temperature on salinity. At salinities below the intersection of the two lines (24.7 °C) an inverse temperature stratification may develop

stratification, despite the strong effect of salinity on density (a salinity difference of 0.14 has the same effect on density as a temperature difference of 1 K). Therefore, the most important function of the vertical salinity distribution lies in the formation of the permanent halocline.

Finally, the density anomaly, a peculiarity of brackish water systems that is related to the dependency of density on temperature, should be addressed. Freshwater has its maximum density at a temperature of T_{max} = 3.98 °C, but this characteristic temperature decreases with increasing salinity. At a salinity of 24.7, T_{max} is identical to the freezing point (Fig. 3.4), which means that at higher salinities the theoretical decrease in density with temperature has no practical meaning because of the formation of ice. However, because of the density anomaly, at salinities <24.7, a surface-water layer colder than the underlying water, and thus an inverse stratification, may be formed during cooling in winter. These conditions mostly occur at low salinities and during the strong winter cooling, hence, in the northern regions of the Baltic Sea.

3.3 Stagnation and Inflow Events

Another hydrographically and biogeochemically important feature of the Baltic Sea is the occurrence of anoxia in the deep waters of its major basins. It is a consequence of the permanent halocline and the topography of the Baltic Sea Proper. The contours of the sea floor along a line between the southwest Baltic Sea and the northern boundary of the Gotland Sea, the largest deep basin of the Baltic Sea, are shown in Fig. 3.5. The deep waters of the individual basins are separated from the surface waters by a permanent halocline that is permeable enough to maintain a certain salinity of the outflowing surface waters (Figs. 3.1 and 3.5) but prevents the delivery of an effective oxygen supply from the surface to the deep waters. Since oxygen consumption in the deep water is faster than the oxygen flux across the halocline, oxygen is depleted and hydrogen sulfide forms, as long as lateral transport does not cause water renewal. Those time spans are called stagnation periods.

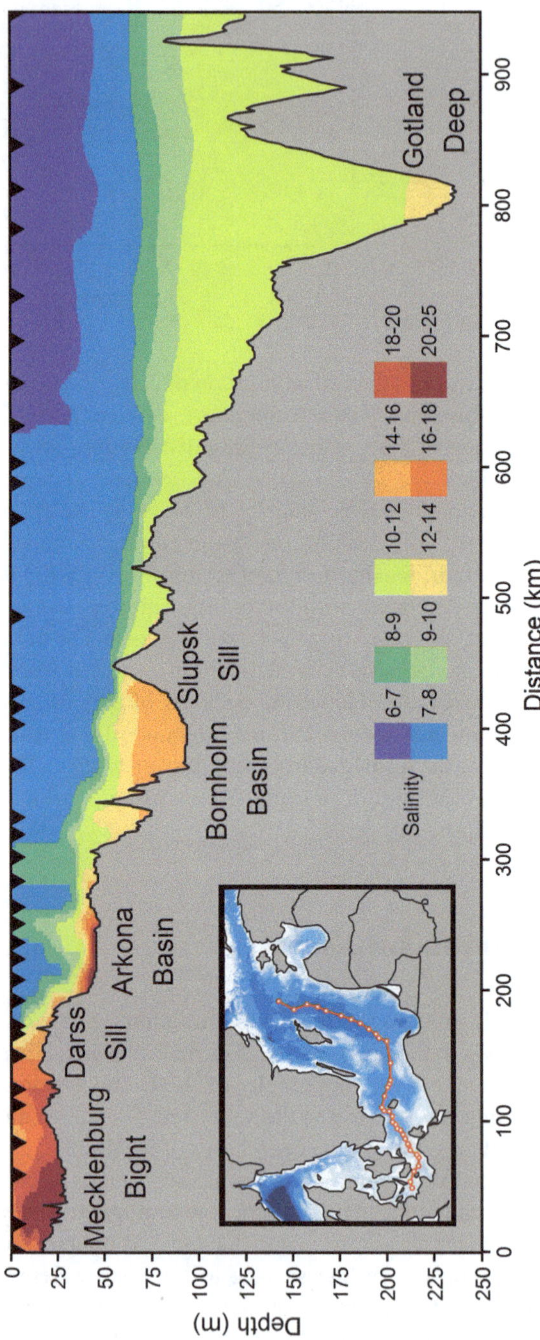

Fig. 3.5 Bottom topography and salinity stratification along a cross section (see inset) from the Mecklenburg Bight to the central Gotland Basin (IOW Monitoring Data, 2013)

However, water renewal by lateral inflow is hampered by the sills that separate the individual basins and takes place only under specific meteorological conditions. Strong inflow events are triggered by a lowered sea level in the Baltic Sea proper that may develop during periods of persistent easterly winds. Upon a subsequent change to strong northwesterly winds, the oxygen-rich water masses piled up in the Kattegat are pushed into the Baltic Sea Proper. Depending on the volume and salinity (density), the inflowing water may even enter the deep basins, marking the beginning of a new stagnation period. The frequency of inflow events is irregular: stagnation periods may be as short as only a few years but also as long as 10–20 years (Mohrholz et al. 2015).

Comprehensive presentations of the Baltic Sea hydrography are given in Feistel et al. (2008) and in Leppäranta and Myrberg (2009).

References

Feistel R, Nausch G, Wasmund N (eds) (2008) State and evolution of the Baltic Sea, 1952–2005: a detailed 50-year survey of meteorology and climate, physics, chemistry, biology, and marine environment. Wiley-Interscience, Hoboken

HELCOM (2006) Development of tools for assessment of eutrophication in the Baltic Sea. In: Balt Sea environment proceedings no. 104

Hjalmarsson S, Wesslander K, Anderson LG, Omstedt A, Perttila M, Mintrop L (2008) Distribution, long-term development and mass balance calculation of total alkalinity in the Baltic Sea. Cont Shelf Res 28:593–601

Leppäranta M, Myrberg K (2009) Physical oceanography of the Baltic Sea. Springer, Berlin

Mohrholz V, Naumann M, Nausch G, Krüger S, Gräwe U (2015) Fresh oxygen for the Baltic Sea—an exceptional saline inflow after a decade of stagnation. J Mar Syst 148:152–166

Chapter 4
The Database

4.1 Studies of the Surface Water CO_2 System

The data in the following sections were used to unravel the biogeochemistry of organic matter (OM) production and mineralization and they originate mainly from research and long-term observational programs at the IOW. The control of the intensity and timing of OM production was of central interest when systematic investigations of the marine CO_2 system were started at the IOW in the 1990s. Measurements of the CO_2 partial pressure (pCO_2) and total CO_2 concentrations (C_T) were performed in connection with the Baltic Sea monitoring program and provided the first insight into the seasonality of the surface-water CO_2 system. The seasonal patterns evident in the data obtained from five measurement campaigns during 1995–1998 (Fig. 4.1) clearly indicated the dominating role of OM production in the seasonality of the pCO_2 in the Baltic Sea, since pCO_2 values were lowest during the productive period (May–July), when the elevated temperature would have caused a pCO_2 maximum in abiotic seawater. Detailed information on the relationship between OM production and the status of the CO_2 system was obtained in 2001, through the project CARFIX. The surface-water pCO_2 distribution in the eastern Gotland Sea was measured at monthly intervals from April to September. The data revealed an unexpected bimodal distribution of the pCO_2, with minima in May and July (Fig. 4.1) that were attributed to OM production during the spring bloom and the mid-summer nitrogen (N) fixation period. The data facilitated estimates of the corresponding production rates and an assessment of the role of nutrient availability, including N-fixation (Schneider et al. 2003). Still, many questions remained unanswered and new ones arose. Their answers necessitated more precise quantification and dating of the seasonal amplitudes and a determination of the interannual and regional variability in the seasonal pCO_2 patterns to fully exploit the potential of pCO_2 measurements for production studies. However, this in turn required a long-term and area-wide pCO_2 time series of increased temporal resolution. This could not be achieved by the use of research vessels, because of the high costs and

B. Schneider and J.D. Müller, *Biogeochemical Transformations in the Baltic Sea*, Springer Oceanography, DOI 10.1007/978-3-319-61699-5_4

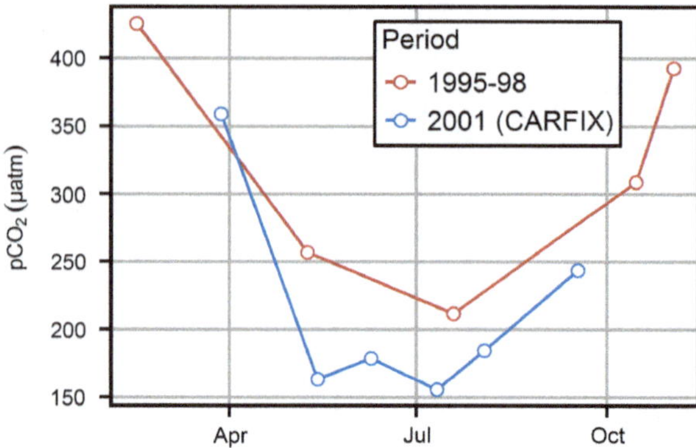

Fig. 4.1 Early representations of the pCO_2 seasonality in surface water of the eastern Gotland based on measurements during research and monitoring cruises

limited availability of ship time. The only alternative was to deploy an automated measurement system on a voluntary observing ship (VOS).

The realization of this idea was fostered by the willingness for cooperation of the scientists at the former Finnish Institute for Marine Research (FIMR), who for several years had made use of VOS lines in the Baltic Sea to record chlorophyll fluorescence, temperature, and salinity (Algaline Project). Using this infrastructure, a fully automated pCO_2 measurement system was installed in 2003 on the Finnlines cargo ship "Finnpartner," which commuted regularly between Helsinki and Lübeck.

Fig. 4.2 Voluntary observation ship (VOS) "Finnmaid"

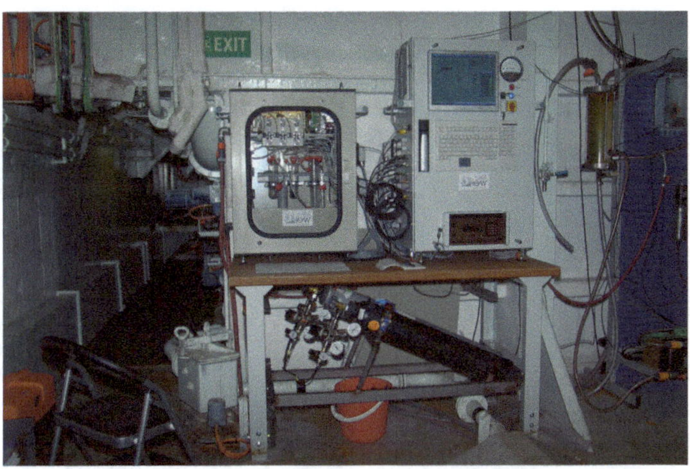

Fig. 4.3 Automated measurement system on VOS "Finnmaid" for the recording of the surface water pCO_2

The automation included regular calibration at the beginning of each transect and a shutdown of the system in the vicinity of polluted harbor waters. pCO_2 was determined by measuring the CO_2 content of air that had passed in a closed loop surface water pumped continuously into a so called bubble-type equilibrator (Fig. 4.3, Schneider et al. 2006). The CO_2 content of the equilibrated air was detected by infrared spectroscopy. The equilibration time was ~5 min and corresponded to a spatial resolution of the data of 1–2 nautical miles at the ship's speed of 20 knots. In 2005, an optode for recording the surface-water oxygen concentration was added to the measurement system. However, only oxygen data collected since 2012 are used in the following discussion, because of difficulties with the calibration of the optode in the years before.

The measurements were substantially extended in 2009, when a second, independent equilibrator and a cavity ring-down spectroscopy (CRDS)-based system were installed that allowed determination of the methane partial pressure. This system also recorded the pCO_2 and comparison of the two independently obtained pCO_2 datasets facilitated the quality control of the measurements. A mean difference between individual transects of <3 µatm was considered to be satisfactory; differences of 3–5 µatm were still accepted but demonstrated the need for careful supervision of the performance of either measurement system. In general, the data from the original system (infrared spectroscopy) were used for further evaluation of the data, except when clear evidence existed for higher quality of the CRDS data.

In 2006, the Helsinki–Lübeck line was taken over by the cargo ship "Finnmaid" The setup of the infrastructure, including re-installation of the equipment on the new ship, caused a gap in our pCO_2 time series of more than a year; thus, the data obtained during 2006 and 2007 were excluded from the biogeochemical analysis. In addition to this larger break, short-term interruptions of the measurements occurred

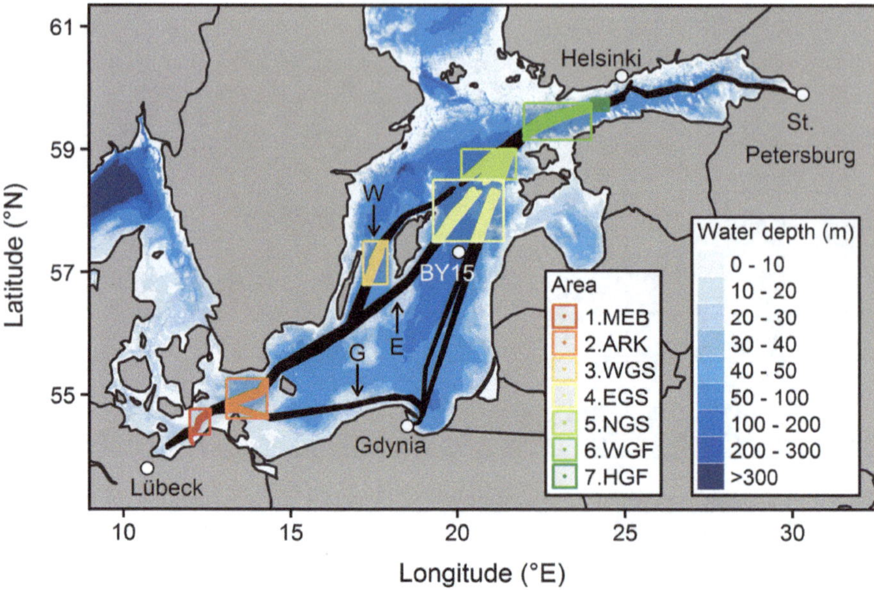

Fig. 4.4 Map showing the different routes, east (E) and west (W) of Gotland and with a harbour stop in Gydnia (G), of VOS "Finnmaid" (before 2007 "Finnpartner"). Seven sub-transects were defined and the respective mean pCO_2, temperature and salinity data from each crossing were used for further data analysis: *MEB* Mecklenburg Bight, *ARK* Arkona Sea, *WGS* Western Gotland Sea, *EGS* Eastern Gotland Sea, *NGS* Northern Gotland Sea, *WGF* Western Gulf of Finland and *HGF* Gulf of Finland, approach to Helsinki. BY15 is a standard monitoring station in the central Gotland Sea

due to occasional malfunctions of any component of the complex measurement system and to dock stays of the ship.

"Finnpartner" and, later on, "Finnmaid" took different routes when traveling between Helsinki and Lübeck (Fig. 4.4). In most cases the ships passed the eastern Gotland Sea, but they occasionally took the route through the western Gotland Sea, mainly when the seas were heavy. In addition, every second month during 2009–2012 Finnmaid made a harbor stop in Gdynia and beginning in 2012 also occasionally visited St. Petersburg. Almost 1600 pCO_2 transects between Helsinki and Lübeck obtained during 2003–2015 provide the database used to analyze the biogeochemistry of OM production in the Baltic Proper. Since the Baltic Proper is characterized by pronounced southwest-northeast gradients both for salinity and biogeochemical variables such as nutrient concentrations, the route of the cargo ship was subdivided into seven sub-transects (Fig. 4.4). These followed the sequence of basins between the Mecklenburg Bight and the western Gulf of Finland, avoiding shallow coastal waters because they may have been affected by direct interactions with the sediment surface or affected by upwelling events throughout the year. The frequency of data recording in the individual sub-transects is shown in Fig. 4.5.

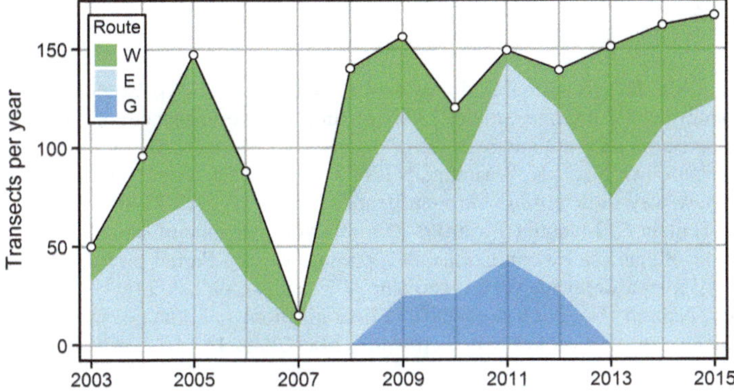

Fig. 4.5 Graphical representation of the number of sub-transects per year that were obtained for MEB, ARK, NGS, WGF and HGF (visited on all routes, *white points*), for EGS (visited on routes E & G, *light and dark blue areas*) and for WGS (visited only on route W, *green area*). The routes E, W, and G and sub-transect locations are shown in Fig. 4.4

4.2 Investigations of the Deep Water CO_2 Accumulation

Along with OM production, which can be quantified from the surface-water pCO_2 measurements, OM mineralization is of particular interest because it leads to oxygen depletion and eventually to hydrogen sulfide formation in deep water layers. The Gotland Basin is an ideal location to study the kinetics and stoichiometry of the various oxidation processes driving the OM mineralization. Due to long-lasting stagnant conditions especially below 150 m, the basin can be considered as a test tube where mineralization takes place continuously and mineralization products accumulate as long as OM is replenished. In contrast to the biogeochemical cycles in the surface water, deep-water OM mineralization shows a less pronounced seasonality (Schneider et al. 2010). A "reset" of the conditions occurs only when a stagnation period is terminated by a water renewal event due to the lateral input of younger water masses. Hence, CO_2 as the primary mineralization product may accumulate over years, and total CO_2 concentrations directly reflect the progress of mineralization. Interfering processes such as $CaCO_3$ formation or dissolution are negligible in the deep waters of the Baltic Proper. The vertical distribution of C_T was therefore determined below the permanent halocline at the central Gotland Sea monitoring station BY15 (Fig. 4.4), using the high-precision/accuracy coulometric titration method (Johnson et al. 1993). Measurements were made at intervals of 2–3 months beginning in 2003 and in conjunction with the IOW long-term hydrographic and hydrochemical observation program. Hence, data for all other relevant biogeochemical and hydrographic variables were available.

References

Johnson KM, Wills KD, Buttler DB, Johnson WK, Wong CS (1993) Coulometric total carbon dioxide analysis for marine studies: maximizing the performance of an automated gas extraction system and coulometric detector. Mar Chem 44:167–187

Schneider B, Nausch G, Nagel K, Wasmund N (2003) The surface water CO_2 budget for the Baltic Proper: a new way to determine nitrogen fixation. J Mar Syst 42:53–64

Schneider B, Kaitala S, Maunula P (2006) Identification and quantification of plankton bloom events in the Baltic Sea by continuous CO_2 partial pressure Partial pressure and Chlorophyll achlorophyll a measurements on a cargo ship. J Mar Syst 59:238–248

Schneider B, Nausch G, Pohl C (2010) Mineralization of organic matter and nitrogen transformations in the Gotland Sea deep water. Mar Chem 119:153–161

Chapter 5
Surface Water Biogeochemistry as Derived from pCO_2 Observations

5.1 Seasonal and Regional Patterns of pCO_2 and C_T

5.1.1 Characteristics of the pCO_2 Time Series

To illustrate the temporal and regional coverage of the pCO_2 measurements, the data from each of the transects are displayed in Fig. 5.1 as a function of time and latitude (Hovmöller diagram). The white vertical stripes mark periods during which no or erroneous measurements were made. The horizontal white stripes indicate the latitudes where the recording of the surface-water pCO_2 was interrupted to measure the CO_2 concentration of the ambient air in the measurement room on the ship. However, as only the air data collected on "Finnpartner" (2003–2006) were unaffected by contamination only they were used to characterize the seasonality of the atmospheric CO_2 over the Baltic Sea (see Sect. 2.1 and Schneider et al. 2014a). The variation in the color pattern reveals a distinct seasonality of the pCO_2 that was more pronounced in the northern than in the southern Baltic Proper. This feature is displayed in greater detail in the pCO_2 time series plots for the mean pCO_2 for each crossing of the individual sub-transects (Fig. 5.2). Due to the low data coverage in the Western Gotland Sea (WGS), where the ship passed only occasionally, this sub-transect was excluded from most of the following analysis.

For the Eastern Gotland Sea (EGS), the pCO_2 data are complemented by time-series data for the surface-water temperature (Fig. 5.3), which show a distinct inverse correlation with the pCO_2 seasonality. pCO_2 minima considerably below the atmospheric level were recorded during the spring/summer temperature maximum in all sub-transects. This contrasts with the thermodynamic temperature dependency of the pCO_2 and indicates that CO_2 consumption during spring/summer biomass production was the major controlling factor in the development of the pCO_2 minima (see Sect. 2.3.4). Since the surface water was strongly undersaturated during this period, the uptake of atmospheric CO_2 counteracted CO_2 depletion by biomass production. However, because gas exchange is much slower than

B. Schneider and J.D. Müller, *Biogeochemical Transformations in the Baltic Sea*, Springer Oceanography, DOI 10.1007/978-3-319-61699-5_5

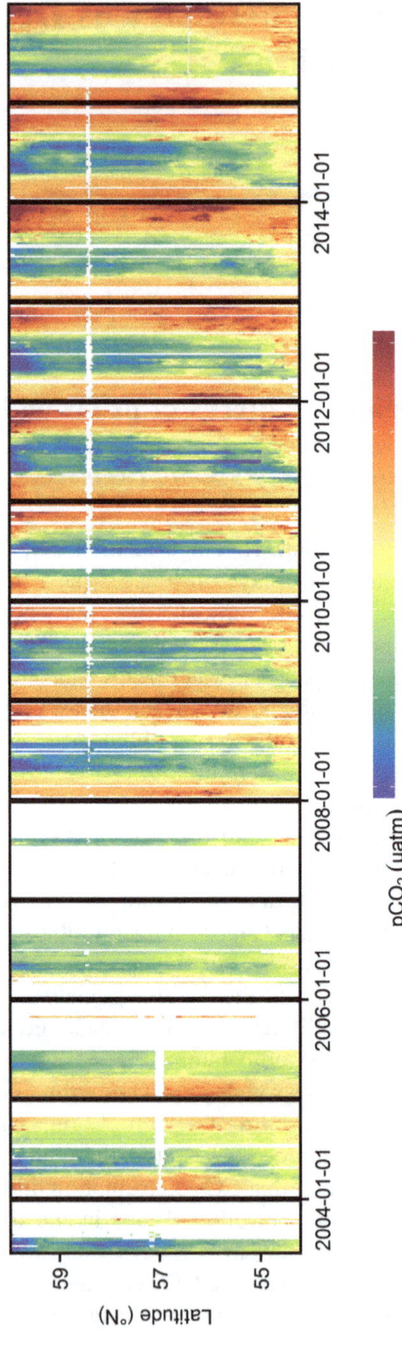

Fig. 5.1 Surface water pCO_2 in the Baltic Sea as a function of time and latitude. The range of the color scale was restricted to 100–600 µatm in order to obtain a reasonable resolution of the pCO_2 seasonality. The gridded data represent weekly mean values and are averaged over 0.02° in latitude

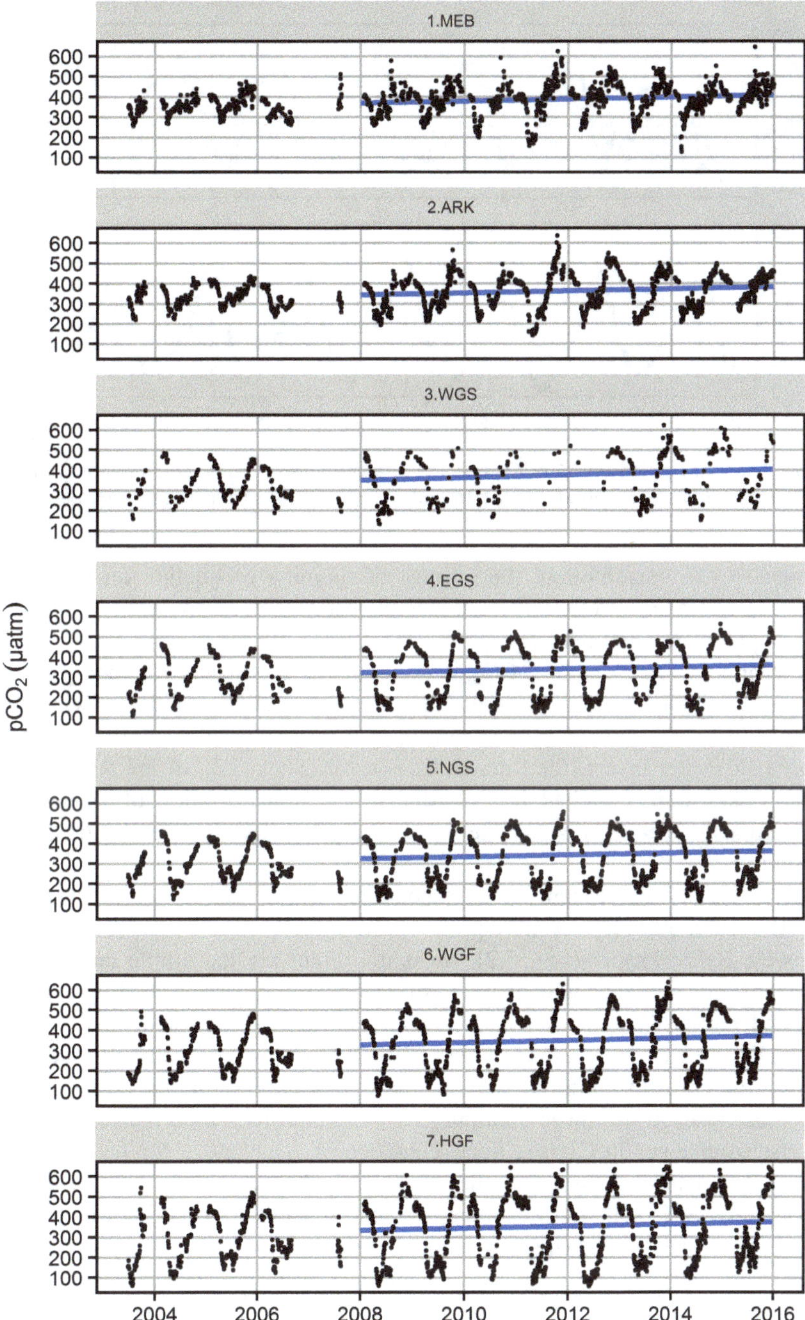

Fig. 5.2 pCO_2 time series for the seven sub-transects defined in Fig. 5.3. The *blue line* represents a linear regression analysis for the period 2008–2015 based on daily interpolated data. Values below 50 µatm and beyond 650 µatm are not shown but included in the regression analysis

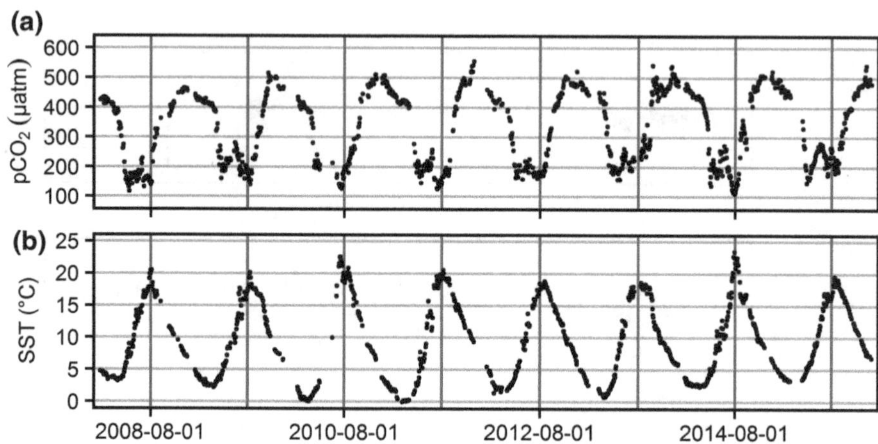

Fig. 5.3 Anti-correlation of the pCO_2 and sea surface temperature in the Northern Gotland Sea (NGS). The *red vertical lines* at August 1st indicate roughly the temperature maximum

biological CO_2 consumption, the imprint of biomass production on the pCO_2 is conserved for many weeks and facilitates the estimation of production rates on the basis of pCO_2 measurements.

A strong increase in the pCO_2 occurred after the spring/summer productive period and coincided with erosion of the seasonal thermocline and with wind-driven vertical mixing down to the permanent halocline. This resulted in the transport of deeper cold water layers, the intermediate water (Sect. 3.2), to the surface. These water masses were enriched in CO_2 due to the mineralization of OM and they generated a surface-water pCO_2 that in some of sub-transects occasionally exceeded 600 µatm.

For a variety of reasons, the seasonality of the pCO_2 as reflected in the peak-to-peak amplitude was more pronounced in the northeast than in the south and southwest Baltic Proper (Fig. 5.2). Most important are the nitrate and phosphate concentrations gradients, which are major drivers of biomass production and lead to enhanced biomass production in the central Baltic and especially in the western Gulf of Finland. However, even without any differences in biomass production, the spring/summer pCO_2 minima will be more pronounced in the northeast Baltic Proper because of its lower alkalinity, which reinforces the change in pCO_2 in response to changes in C_T (see Sect. 2.3.4).

An additional influence on the seasonal pCO_2 signal is the Baltic Sea's hydrography. Seasonal stratification is especially stable in the central and northern Baltic Proper such that mixing across the thermocline is for the most part hampered. Hence, the depletion of C_T by biomass production in the surface water is widely maintained because the counteracting process, the mineralization of particulate organic matter (POM), takes place mainly below the surface mixed layer. This spatial separation of net CO_2 depletion and CO_2 accumulation is abrogated with the onset of deep mixing in autumn, when the CO_2 that previously accumulated below the thermocline is transported to the surface, thus leading to the autumn/winter pCO_2 maximum. In the southwest Baltic Proper, however, in addition to lower nutrient concentrations and biomass production, stratification of the water column is less stable. The separation of deeper water layers, where mineralization prevails, from the surface mixed layer may therefore be abolished, even during spring and summer. This is a consequence of the dynamic current pattern in the transition area between the Baltic Sea and the North Sea.

5.1.2 *Long-Term Changes in* pCO_2

Since the pCO_2 measurements presented here cover a considerable time span, a simple trend analysis was performed. Unfortunately, the data before 2008 could not be used because of the large gaps, especially during wintertime, when the pCO_2 typically reaches its maximum. Therefore, only data from 2008 to 2015 were included in the trend analysis. All observations in this time period were linearly interpolated to generate daily data and to ensure a homogenous distribution of data points in time. A simple linear regression analysis applied to this interpolated data set yielded positive pCO_2 trends for all sub-transects, ranging between 4.6 and 6.1 μatm $year^{-1}$. All trends are significant, with p-values $\ll 0.001$ and standard errors ranging from 0.6 to 1.5 μatm $year^{-1}$. A striking feature of this simple linear regression analysis are the very low R^2 values, between 0.007 and 0.023, which reflect the large seasonal variability of the surface water pCO_2 in the Baltic Sea, which can be on the order of hundreds of μatm (Figs. 5.1 and 5.2). A comparable linear regression analysis for pCO_2 observations from the Bermuda Atlantic Time Series stations yielded a positive trend of $+1.62 \pm 0.21$ μatm $year^{-1}$ for the time period 1983–2011 (Bates et al. 2012). The corresponding R^2 of 0.16 reflects the far less pronounced seasonal variability in the open North Atlantic Ocean.

Whereas the pCO_2 trend in North Atlantic waters corresponded approximately to the atmospheric CO_2 increase, this was not the case for the Baltic Sea, where the trends considerably exceeded the atmospheric CO_2 increase. The reasons for this may be manifold and are a consequence of the complex hydrological and biogeochemical dynamics. pCO_2 evolution in the Baltic Sea might on decadal scales be impacted by the interplay between inflow events and stagnation periods. The first 7 years of the analyzed time period were part of a deep-water stagnation period that lasted from 2004 to 2014. During this time, POM produced in the surface waters sank to the sediment surface, where it was mineralized. Hence, CO_2 was produced and progressively accumulated in deeper water layers (Schneider et al. 2010). Due to vertical mixing, it slowly migrated upwards and increased the C_T gradient in the vicinity of the permanent halocline. Hence, the input of C_T by mixing into the surface water was enhanced, giving rise to a trend in the winter pCO_2 that was formed during mixing of the surface water down to the halocline.

Furthermore, changes in temperature may cause a trend in pCO_2. Analysing the development of the sea surface temperature according to the same principles applied to the pCO_2 data yields positive trends for all sub-transects in the range of 0.17–0.24 K $year^{-1}$ (standard error: 0.04–0.05 K $year^{-1}$, $p \ll 0.001$, $R^2 < 0.01$) for the period 2008–2015. This might also have contributed to the increase in pCO_2, however, quantification of this effect is not trivial since it must account for the mitigation of the pCO_2 increase by gas exchange.

Finally, elevated surface water pCO_2 levels may also increase the seasonal pCO_2 amplitude for a given change in C_T caused by OM production or mineralization. This is a consequence of the increasing R* (= $dpCO_2/dC_T$) with increasing pCO_2 (Fig. 2.10). To illustrate this effect, we consider the example of a spring bloom in 2008, in which production started at a pCO_2 of 386 μatm and the total C_T consumption amounted to 50 μmol kg^{-1}. At a temperature of 5 °C and an alkalinity typical for the central Baltic Sea of 1650 μmol kg^{-1} (salinity = 7), this would have caused a drawdown of the pCO_2 by 200 μatm. Seven years later, in 2015, when the increase in atmospheric CO_2 elevated the surface-water pCO_2 by 14 μatm, the same C_T consumption during the spring bloom would have caused a 209 μatm drop in the pCO_2. Hence the seasonal drawdown of pCO_2 increased by about 1.3 μatm per year.

5.1.3 The Seasonal Fine Structure of the pCO_2

For a more detailed assessment and interpretation of the seasonality of pCO_2, all pCO_2 data from 2003 to 2015 were plotted versus the day of the year for each of the individual sub-transects (Fig. 5.4). The data refer to the individual years and were not normalized for the annual increase of atmospheric CO_2 indicated by the thickness of the gray bands in the figure. They were ordered according to their location between the Mecklenburg Bight (MEB) and Gulf of Finland. Before discussing regional differences, we present an interpretation of the seasonal fine structure of the pCO_2 in the northern Gotland Sea (NGS) (Fig. 5.4). This sub-transect best represented open Baltic Sea conditions while providing high temporal data coverage, since it was visited during both the western and the eastern routes of the ship (Fig. 4.4). The occurrence of two distinct pCO_2 minima is the most prominent feature of the seasonal distribution patterns. On average, the first minimum occurred in the beginning of May and reflected CO_2 consumption by the spring bloom. The values in some years were as low as 100 μatm; since these are the mean values within the sub-transect, this implies that some of the single values of the 1-min record were distinctly <100 μatm. This applies also to the second minimum, which appeared in early July during almost all years of the study period and reflected mid-summer production fuelled by nitrogen fixation. It must be noted that the decrease in the pCO_2 is not proportionally related to the intensity of the production events, because pCO_2 is also a function of temperature and because a drop in the pCO_2 upon a change in C_T is more pronounced at high pCO_2 levels (see Sect. 2.3.4). The time span between two CO_2 minima is sometimes referred to as the “blue water” period because net production is low and chlorophyll-a

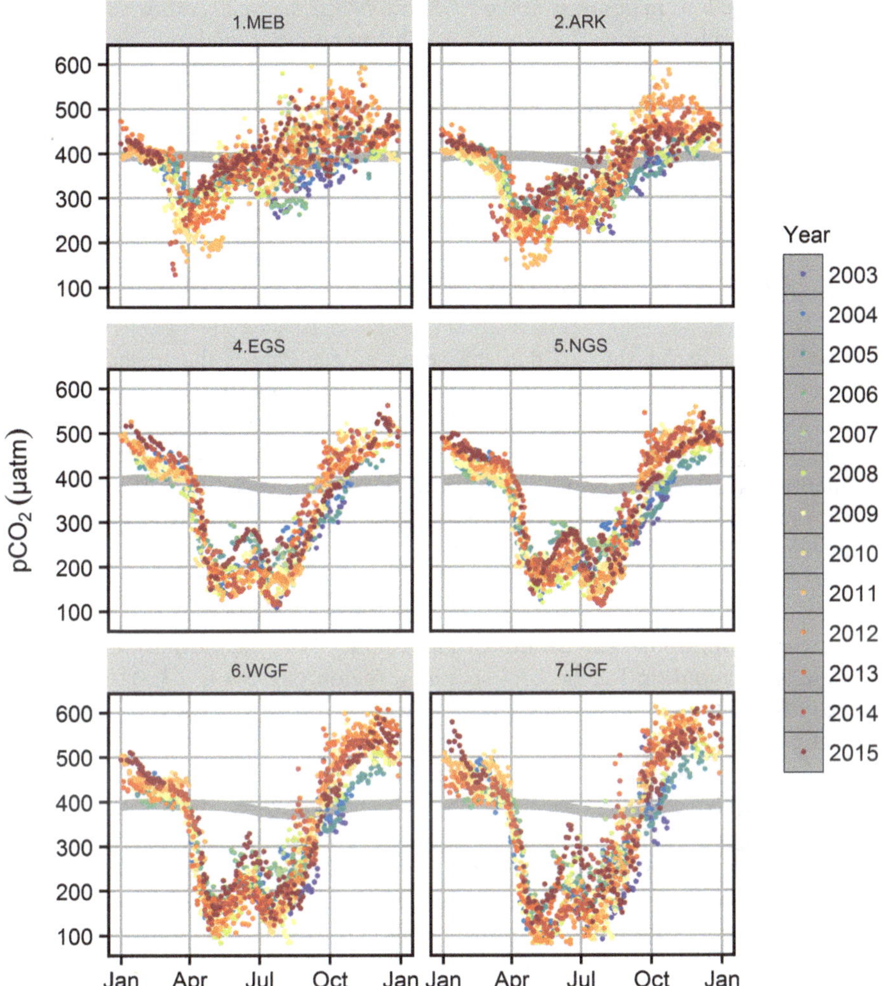

Fig. 5.4 The pCO_2 data as a function of the date of the year, plotted separately for all data from the individual sub-transects. The color refers to the individual years. The *grey* band shows the atmospheric pCO_2 and its width represents the range of atmospheric pCO_2 increase during 2003–2015

concentrations are too low to give the water the typical greenish color conferred by the spring bloom. At the same time, rising temperatures and CO_2 uptake from the atmosphere cause a pCO_2 maximum by mid June.

Similar seasonal pCO_2 patterns were observed in the EGS and WGS, although fewer data were available for the latter. Two pCO_2 minima occurred also in the western Gulf of Finland (WGF) and its Helsinki region (HGF). They were characterized by even lower mean spring and mid-summer pCO_2 values, but an enhanced interannual variability. In the Arkona Sea (ARK), bimodal seasonality was still visible but, in contrast to the EGS and NGS, the pCO_2 minimum was less pronounced in mid-summer than in spring. This trend proceeded towards the MEB, where a pCO_2 minimum in July/August was only occasionally determined. This indicated a decrease in the abundance of N-fixing cyanobacteria towards the southwest Baltic Proper, consistent with previous biological observations (Wasmund 1997).

5.1.4 From pCO_2 Measurements to Total CO_2 Data

The above discussion showed that the seasonal patterns of the mean pCO_2 for different sub-transects provide an adequate qualitative description of the annual succession occurring within the main production periods, including its regional peculiarities. However, to quantify production rates, typically by mass-balance calculations, changes in the CO_2 system must be expressed in terms of concentration units. Therefore, total CO_2 concentrations were calculated from the pCO_2 measurements. This requires information on a second variable of the marine CO_2 system, preferably alkalinity, which in our study was not available. However, because we were not interested in an accurate absolute C_T, but in its change over time, ΔC_T, it sufficed to estimate the mean alkalinity from the mean salinity and the A_T-salinity relationship for the corresponding region (see Sect. 2.3.3). The uncertainties associated with the calculated ΔC_T were acceptable, because the change in C_T that corresponds to a change in pCO_2, $\Delta C_T/\Delta pCO_2$, is only a weak function of A_T. Referring to the conditions in the eastern Gotland Sea, an upper limit error in the estimated A_T of ± 50 µmol kg^{-1}, yielded an uncertainty in the calculation of ΔC_T of only ± 5 µmol kg^{-1} (Schneider et al. 2009). However, this approach implies that there is no internal A_T consumption or generation because both would have a strong effect on the pCO_2 that is not caused by changes in C_T. In many pelagic marine environments, the growth of calcifying plankton may significantly increase the pCO_2 and thus counteract the decrease caused by OM production. However, the abundance of such organisms is negligible in the Baltic Sea, except in the Kattegat area (Tyrrell et al. 2008), which was not included in our investigations. We also ignored the small change in A_T (3–5 µmol kg^{-1}) related to nitrate consumption, which is accompanied by the removal of a proton and increases A_T accordingly. Internal A_T sources are present also in deeper hypoxic and anoxic water layers, due to sulfate reduction and denitrification (Gustafsson et al. 2014), but they will affect the surface A_T only in the long run.

Based on these assumptions, we calculated the C_T seasonality for each sub-transect using the mean pCO_2 and the sea surface temperature (SST) for each of the VOS crossings of the sub-transect as well as the overall mean salinity and the resulting overall mean alkalinity for the considered time period. The latter was confined to the productive months of March to August since our main interest was C_T depletion by OM production. C_T was then calculated using the carbonic acid dissociation constants suggested by Millero (2010) and the CO_2 solubility constants

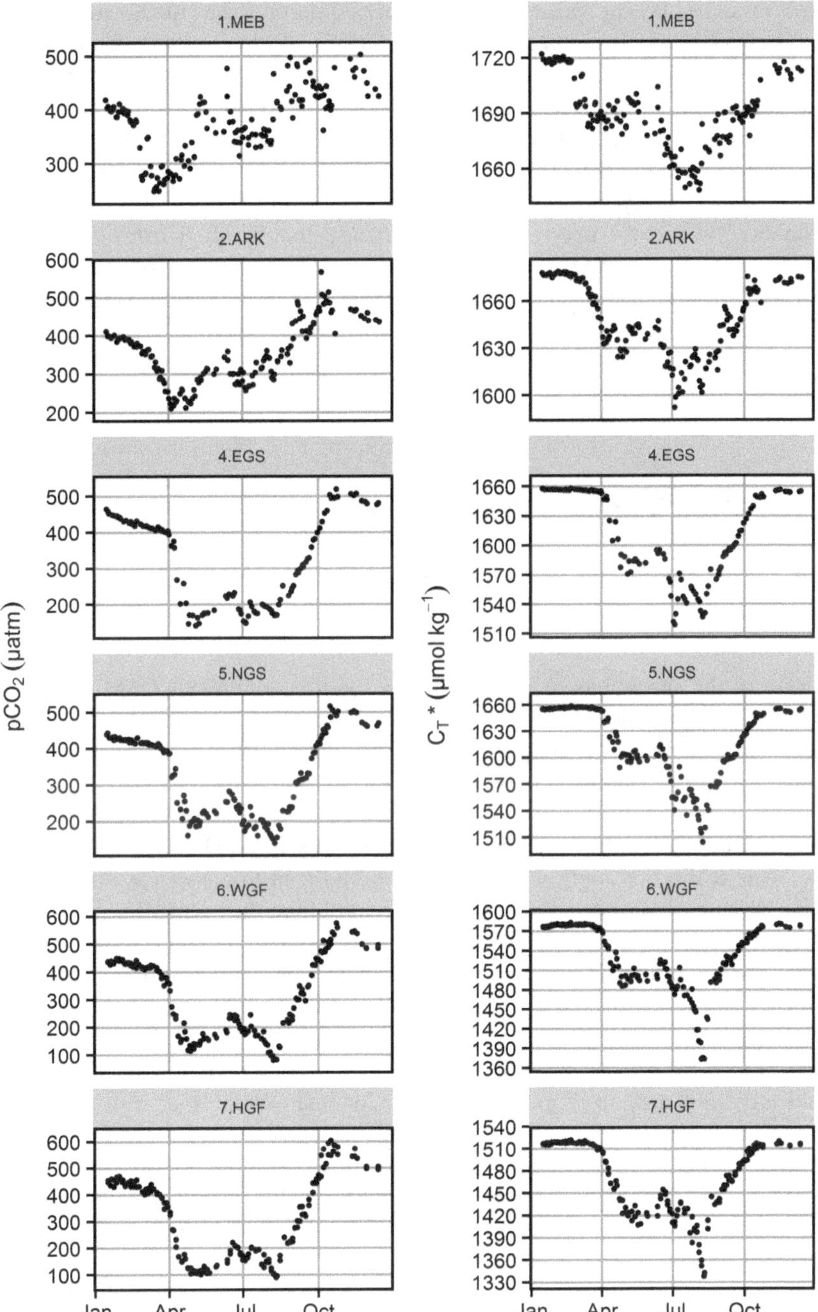

Fig. 5.5 pCO_2 (*left*) and C_T* (*right*) as a function of the date of the year in 2009, plotted separately for the individual sub-transects. The pCO_2 seasonality reveals a W-shape pattern, whereas the C_T* shows rather a stepwise progression, which is more pronounced in the northern regions. Note that the scaling of y-axes is different for all sub-transects, but the distance between the horizontal grid lines is identical

from Weiss (1974). The values obtained can be interpreted as the C_T normalized to the mean A_T: C_T*. In contrast to pCO_2, C_T* is a conservative variable (in the physico-chemical sense, see Sect. 2.3.4) and the seasonality of the calculated C_T * is not affected by changes in temperature. Hence, the distinct intermediate increase in pCO_2 in the central and northern Baltic Proper during late spring, which was mainly due to the temperature increase (see Sect. 5.2.5) and caused a W-shape of the pCO_2 seasonality, was not reflected in the seasonality of C_T*. Rather, C_T* seasonality followed a nearly stepwise decrease that resulted from biomass production during the spring bloom and mid-summer N-fixation period and corresponded to the two minima of the pCO_2 distribution in the central Baltic Sea. This is illustrated in Fig. 5.5, which shows the seasonality of pCO_2 and C_T* for the individual sub-transects in 2009. Detailed interpretations of the 2009 data are presented in the following sections.

5.2 A Walk Through the Seasons

5.2.1 Timing of the Spring Bloom and the Role of Solar Radiation

The start of the spring bloom was defined as the first continuous C_T* decrease after the winter deep mixing of the surface water that exceeded 1 µmol kg^{-1} d^{-1} and could not be explained by gas exchange, because the surface water pCO_2 was close to the atmospheric pCO_2. Based on these criteria, the starting dates could in most cases be unambiguously identified with an uncertainty (±2 days) corresponding to the frequency of the ship's crossing. This refers especially to 2009 (Fig. 5.6) although in some years it did not apply to the MEB and HGF; in the latter case the estimated starting dates are rather the date when the spring bloom has started at the latest. We attribute this to the lateral transport of water masses in regions characterized by strong biogeochemical gradients, such as in the transition area to the North Sea (MEB). But also near-coast processes (HGF), such as direct interactions between the sediment and surface water, may have masked the start of the spring bloom. The start dates of the spring bloom for each sub-transect and year are compiled in Table 5.1 and are graphically depicted in Fig. 5.7, except for the WGS and a few particular sub-transects/years in which data gaps during spring prevented precise dating.

In the northern sub-transects and the Gulf of Finland, the spring bloom started, on average, on almost the same day, between March 27 (HGF) and March 29 (EGS, NGS) (Table 5.1). The average start dates in the MEB and ARK, were considerably earlier, by roughly 4 and 2 weeks, respectively. This gradient in the onset of the spring bloom south of the Gotland Sea was in agreement with biological observations (Wasmund et al. 1998). In addition to the earlier onset of the spring bloom, a high interannual variability of the starting date was determined in the southern sub-transects MEB and ARK, with a standard deviation of 14 and 9 days,

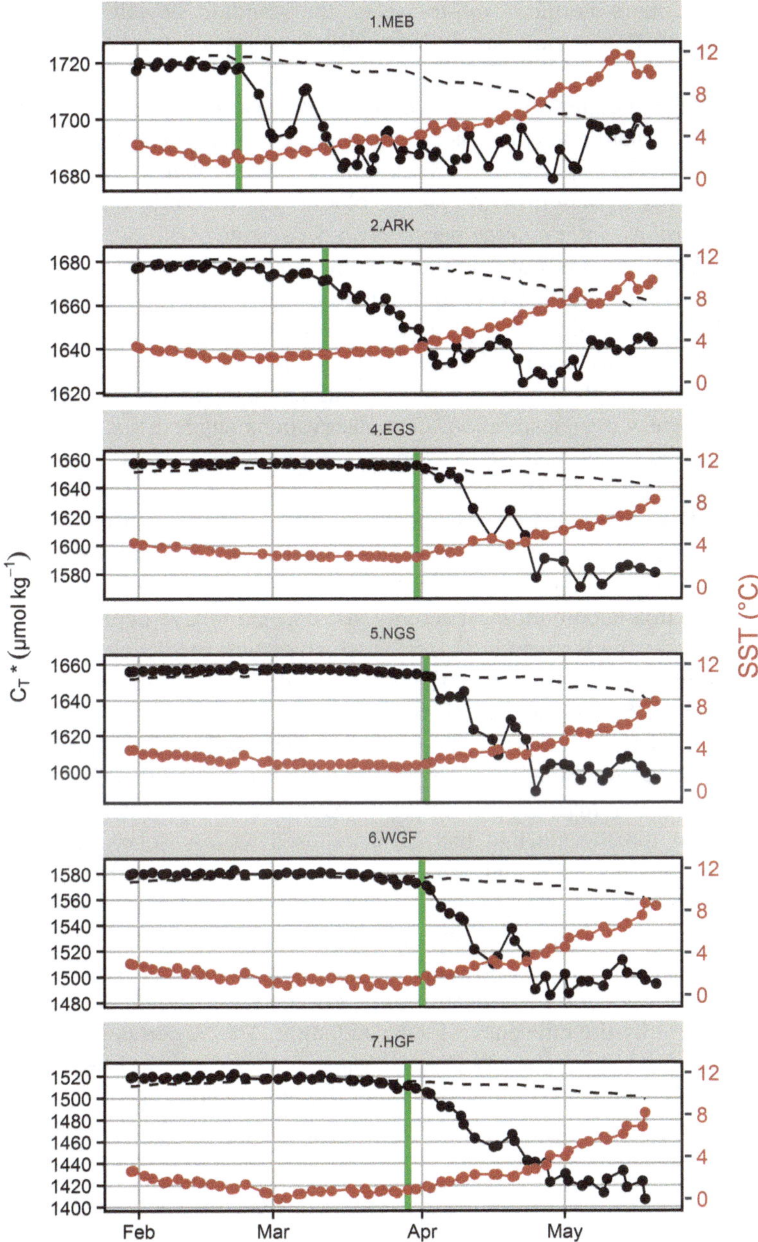

Fig. 5.6 The decrease of C_T* (*black*) during the spring bloom period in 2009 in the individual sub-transects. The *dashed line* indicates the theoretical C_T* for equilibrium with an atmospheric pCO_2 of 400 μatm. The start of the spring bloom is indicated by the *green vertical line*

respectively. In the more northern sub-transects, the standard deviation of the starting dates (Table 5.1) was surprisingly low, only 3–4 days for the EGS, NGS, WGF, and HGF.

These findings raise the question for the control of the start of the spring bloom. Our data clearly show that temperature is not the trigger, as the SST at the starting dates (Fig. 5.7, Table 5.1) varied by several K and indicate that a temperature threshold value for the initiation of the spring bloom was unlikely. Instead, the increase of SST after the winter cooling period always coincided with the depletion of C_T* (Fig. 5.6). Since the onset of warming is a pre-requisite for the formation of the seasonal thermocline, our observations suggest that a stabilization of the spring/summer thermal stratification triggers the development of the spring bloom (Wasmund et al. 1998). Only if vertical mixing does not exceed the so called critical depth (Sverdrup 1953) can phytoplankton absorb sufficient light energy to carry out photosynthesis in excess over respiration. Our observations suggest that the thermocline when detected by an increase in temperature, was already located close to the critical depth.

The coupling of warming and C_T* depletion by OM production in the mixed layer obviously lies in their common cause, solar radiation. Together with the mixed-layer depth, solar radiation controls both short-term fluctuations in seasonal warming and the accumulated biological activity, as reflected in C_T* depletion. An example of this relationship is given in Fig. 5.8a, which depicts the development of SST and C_T* during April 2009 in the NGS. A strong deviation from the general seasonal increase in the SST occurred, for example, from April 17 to 19, when the SST dropped by $\sim$0.4 K. At the same time, the C_T* increased by 20 µmol kg^{-1}. This coincidence can be attributed to a deepening of the thermocline and thus to mixing with deeper water masses. Due to light absorption in the water column, water masses below the thermocline had been exposed to less solar radiation, resulting in their reduced warming and less efficient primary production. Hence, the water masses that mixed into the surface layer on April 17 were colder and less C_T*-depleted. This connectivity between warming and photosynthesis is also documented by the plot of C_T* versus SST obtained using the April 2009 data (Fig. 5.8b). The almost perfect linear relationship between these variables suggests their exclusive control by the efficiency of solar radiation. This is certainly not the case throughout the productive period but applies approximately to the start of the spring bloom, when sufficient nutrients for OM production are available. A discussion of the role of nutrients in the development and productivity of the spring bloom is provided in Sect. 5.2.4.

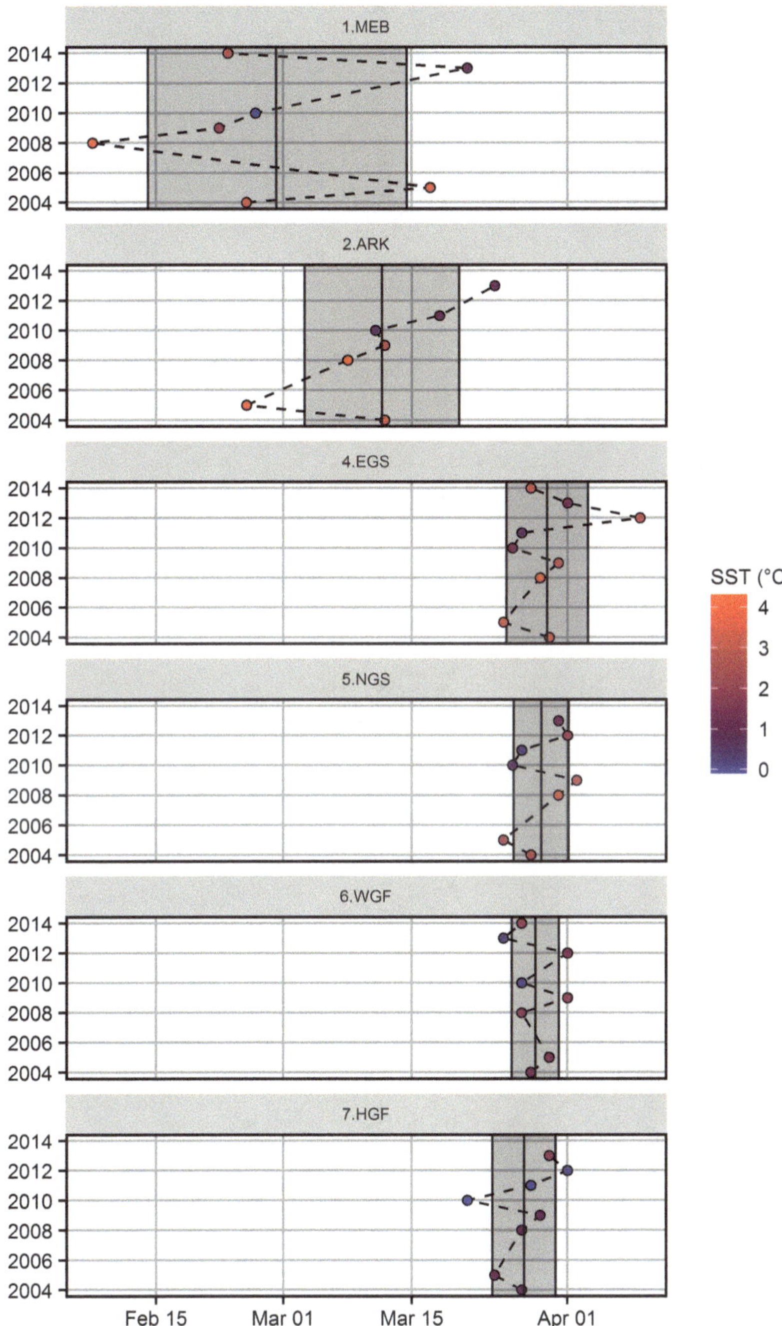

Fig. 5.7 Starting date of the spring bloom for the individual years and sub-transects. The starting date is independent on the sea surface temperature indicated by the color scale. The *vertical lines* represent the mean starting dates for 2004–2014 and grey areas indicate the standard deviations of the starting dates

Table 5.1 Starting date of the spring bloom and corresponding sea surface temperature in °C for the six sub-transects from 2004–2014. A graphical representation of this summary is given in Fig. 5.7

Year	MEB		ARK		EGS		NGS		WGF		HGF	
	Date	SST	Date	SST	Ddate	SST	Date	SST	Date	SST	Date	SST
2004	25-Feb	3.26	11-Mar	3.28	29-Mar	3.17	27-Mar	2.6	27-Mar	1.27	26-Mar	0.88
2005	17-Mar	3.52	25-Feb	3.62	25-Mar	2.98	25-Mar	2.34	30-Mar	1.48	24-Mar	1.11
2008	08-Feb	3.94	07-Mar	4.39	28-Mar	3.59	30-Mar	3.35	26-Mar	1.61	26-Mar	0.9
2009	22-Feb	1.99	12-Mar	2.68	31-Mar	2.84	02-Apr	2.73	01-Apr	1.8	29-Mar	0.89
2010	26-Feb	−0.04	11-Mar	0.59	26-Mar	1.35	26-Mar	0.67	27-Mar	0.08	21-Mar	−0.14
2011	–	–	18-Mar	0.93	27-Mar	0.66	27-Mar	0.2	–	–	28-Mar	−0.11
2012	–	–	–	–	08-Apr	2.68	31-Mar	1.97	31-Mar	1.42	31-Mar	0.07
2013	21-Mar	0.83	24-Mar	0.96	01-Apr	1.43	31-Mar	1.29	25-Mar	0.23	30-Mar	1.83
2014	23-Feb	2.82	–	–	28-Mar	3.05	–	–	27-Mar	1.92	–	–
Mean date	28-Feb		11-Mar		29-Mar		29-Mar		28-Mar		27-Mar	
Std	14		9		4		3		3		3	

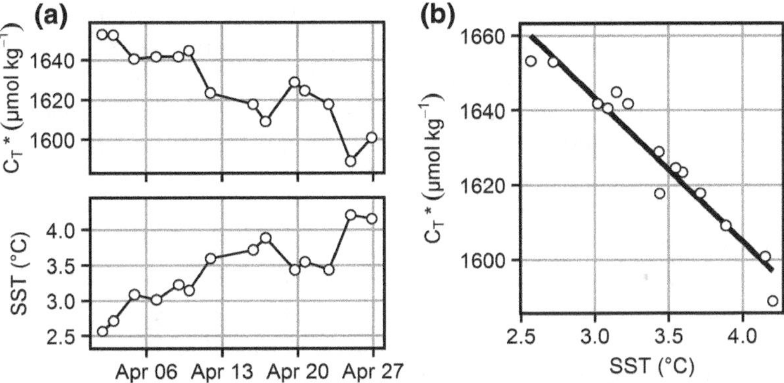

Fig. 5.8 **a** Short-term C_T* peaks superimpose the C_T* decrease during the 2009 spring bloom in the Northern Gotland Sea and coincide with drops in the sea surface temperature caused by mixing events. **b** This is reflected in an almost perfect linear relationship between C_T* and SST

5.2.2 *Quantification of Spring Bloom Productivity*

Definitions of biological productivity abound in the literature. Depending on the biogeochemical implications of the biological production, they include new production, regenerated production, gross production, export production, and net community production (Platt et al. 1984). The latter term applies to our approach for the determination of productivity, which uses mass balance to determine the size of the C_T sink. In the absence of calcifying organisms, an internal C_T loss can only be explained by the OM formation resulting from the interplay between primary production and heterotrophic respiration, referred to as net community production (NCP).

For the following estimates of NCP in the Baltic Sea, the spring bloom, the "blue water" period, and the mid-summer N-fixation period are considered separately. NCP can be quantified in terms of concentration units (mmol-C m^{-3}) and thus describes the density of production. However, if NCP estimates are to be used in assessments of eutrophication and the potential export of OM into deeper water layers, they must be integrated over depth (iNCP) and expressed as mmol-C m^{-2}. In principle, this requires knowledge about the depth distribution of NCP and its interplay with vertical mixing. Since this information is not available from the measurements made on the cargo ship, different approximations have been used herein and in previous studies (Schneider et al. 2006, 2014b).

In the following, the pCO_2 data from 2009 are presented to illustrate the mass-balance calculations used in estimating NCP and to examine the associated problems. During 2009, the pCO_2 time series was hardly affected by either technical failure of the instrumentation or breaks in the ship`s operation. The NCP was calculated for the individual sub-transects except the WGS, where the temporal resolution of the data was too low to allow for a reasonable estimate.

The start of the spring bloom NCP was marked by the first significant decrease in C_T* after the accumulation of C_T* and nutrients during the winter, and its termination, no later than the end of May, by the absence of a further decrease C_T*. However, the data (Figs. 5.6 and 5.8) show that short-term fluctuations were superimposed on the general decrease in C_T*; these must be discussed before presenting the mass-balance calculations. Especially in the MEB, C_T* showed a considerable temporal variability, attributable mainly to the dynamic surface currents that transported water masses of different origin, including near-coastal waters, to the MEB sub-transect. This also accounts for the sudden and strong increase in C_T* that occurred shortly after the start of the spring bloom in the MEB, between March 5 and 7. Figure 5.9 shows the corresponding pCO_2 data along the MEB sub-transect and illustrates how the front between water masses with high and low pCO_2 was moving southwestwards within the MEB longitudinal interval. This lateral transport caused a rise in the mean pCO_2 by 65 µatm, corresponding to an increase of C_T* by 14 µmol kg^{-1}, within a few days in the region from 12.0°E to 12.6°E (MEB). Although C_T* returned to the previous level after a few days, reliable NCP estimates were complicated by these circumstances. This demonstrates a more general problem related to the limits of mass-balance calculations for the constituents of a characteristic water mass. Commonly these calculations are based on concentration measurements that allow the determination of inventories and the temporal changes within a defined compartment. In many cases, additional data are available that facilitate the calculation of fluxes across the vertical boundaries. However, in general, it is extremely difficult to account for lateral transports and their importance for the mass-balance calculations is therefore simply ignored. This deficiency becomes critical if the compartment is of comparatively small dimensions and located in a region with strong gradients and a dynamic current field, as is the case in the MEB sub-region.

Deviations from a continuous C_T* decrease also characterized the other sub-transects in 2009 (Fig. 5.6) and were accompanied by a decrease in SST that

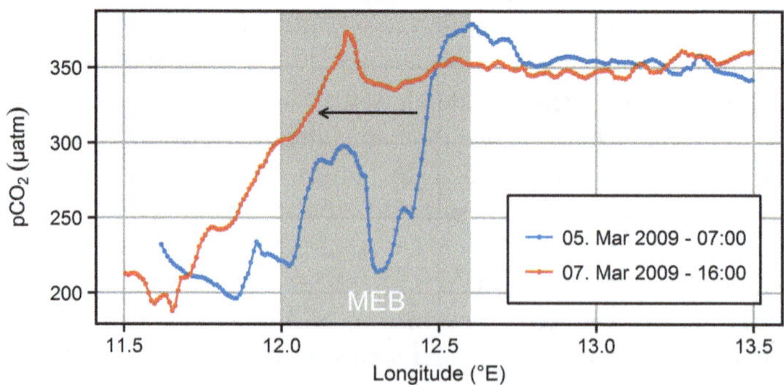

Fig. 5.9 Cause for short-term pCO_2 peak in the Mecklenburg Bight (MEB) during the 2009 spring bloom (see Fig. 5.6). Surface currents (direction indicated by the *arrow*) transport water masses with high pCO_2 into the MEB sub-transect (*grey*) and thereby increase the mean pCO_2 from March 5 (*blue*) to March 7 (*red*)

followed its rapid increase (Fig. 5.8). As discussed in Sect. 5.2.1, this coincidence could be attributed to a wind-driven increase of the mixed-layer depth and is consistent with the sudden C_T* increase occurring first in the ARK (April 17) and then, roughly 2 days later (April 19), in the northern sub-transects EGS, NGS, WGF, and HGF (Fig. 5.6). This delay corresponds approximately to the time needed by a low-pressure area to move eastwards across the Baltic Proper. An example that more directly indicates the role of wind speed in the fluctuations of C_T* during the spring bloom is given in Fig. 5.10. C_T* in the NGS was plotted together with wind speed at a coastal station at the northern tip of Gotland Island using data for the weeks before and after the C_T* peak. The plot shows the appearance of C_T* minima after periods of low wind speeds.

The determination of NCP during the spring bloom was based on a total CO_2 mass balance for a surface-water compartment. Since in most cases the influence of lateral C_T transports can be neglected during the relative short spring bloom period, we considered a one-dimensional compartment (only vertical fluxes were considered and inventories were expressed in moles per area). The upper limit of this compartment is, of course, the air-sea interface, but the definition of the lower compartment boundary is more complex. Previous production estimates used the mean mixed-layer depth to calculate mass balance (Schneider et al. 2006).

However, since the mixed-layer depth varies due to changing wind conditions and photosynthesis is not confined to the mixed layer, the depth at which no further signs of NCP are detectable, z_{pro}, can be used as the lower compartment boundary in the determination of the integrated NCP. To calculate, z_{pro}, we used the monthly monitoring data for the vertical distribution of the phosphate (commonly depicted as PO_4^{3-} although it mainly exists as HPO_4^{2-} at a seawater pH around 8) concentration at 11 stations between the MEB and HGF (provided by SMHI and SYKE).

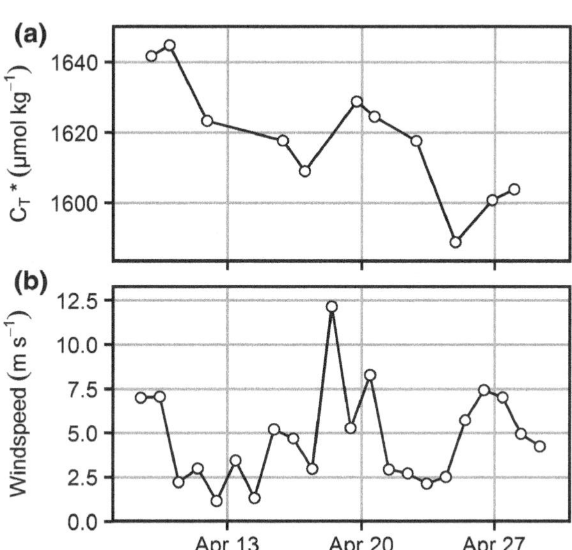

Fig. 5.10 C_T* **a** and wind speed **b** during the spring bloom of April 2009 in the Northern Gotland Sea (NGS). Increasing surface water C_T* levels are caused by wind-induced vertical mixing processes

Table 5.2 Summary of the lower box boundary (z_{pro}), the effective NCP penetration depth (z_{eff}), the C_T depletion (ΔC_T*), the air-sea gas exchange (F_{AS}), as well as the integrated (over depth) net community production (iNCP) and the net community production in concentration units (NCP) for the 2009 spring bloom

sub-transects	z_{pro} (m)	z_{eff} (m)	ΔC_T* (µmol kg^{-1})	F_{AS} (mmol m^{-2})	iNCP (mmol m^{-2})	NCP (µmol dm^{-3})
MEB	21	17	33	489	837	50
ARK	41	35	33	632	1414	41
EGS	52	39	72	554	2709	69
NGS	52	39	55	545	2150	55
WGF	45	26	73	630	1992	78
HGF	38	28	93	726	2645	96

The depletion of PO_4^{3-} at different depth levels after the spring bloom compared to the homogenous distribution of PO_4^{3-} at the start of the spring bloom served as a proxy for the vertical spreading of biological activity and was used to calculate z_{pro} (Box 5.1, Table 5.2). Dividing the integrated PO_4^{3-} loss by the change in the PO_4^{3-} concentration at the immediate surface yielded the effective NCP penetration depth, z_{eff} (Table 5.2), which was used to estimate integral C_T* depletion, iΔC_T*, by multiplication with the observed ΔC_T* in the surface layer (Eq. 5.1 and see Box 5.1):

$$i\Delta C_T^* = z_{eff} \cdot \Delta C_T^* \cdot \delta \tag{5.1}$$

ΔC_T* was obtained from the difference between the C_T* at the start of the spring bloom and the mean C_T* determined at the end of May (May 20–31)(red lines and black bars in Fig. 5.11). Multiplication with the density, δ, is necessary to convert C_T* from a mass- to a volume-related concentration unit, though numerically irrelevant. To estimate the integrated NCP (iNCP$_t$), the air-sea CO_2 gas exchange (F_{AS}) partly compensating for the biogenic loss of C_T* must be taken into account (Eq. 5.2):

$$iNCP_t = z_{eff} \cdot \Delta C_T^* \cdot \delta + \int_{t,start}^{t,end} F_{AS} \cdot dt \tag{5.2}$$

where “i” and subscript “t” indicate that the NCP is integrated over depth and refers to both the particulate and dissolved organic matter, respectively;

Box 5.1: Vertical extrapolation of surface properties
NCP was estimated based on the total CO_2 mass balance for a one-dimensional surface-water compartment whose upper limit is the air-sea interface. The lower compartment boundary, z_{pro}, is defined by the depth at which NCP is no longer detectable. Phosphate (PO_4^{3-}) depletion is used as a proxy for the vertical distribution of NCP. Monthly PO_4^{3-} monitoring data were available with a depth resolution of 5–10 m (Monitoring data provided by SMHI and SYKE) at several stations in the Baltic Proper and the Gulf of Finland. Figure B5.1 provides an example using data from station BY29, in the northern Gotland Sea, and shows that the winter PO_4^{3-} concentrations were homogenously distributed between the surface and the halocline. They were used as reference in the calculation of PO_4^{3-} depletion during the spring bloom. The depth at which the PO_4^{3-} concentration after the spring bloom in mid-May was identical with that of the winter profile was determined by linear interpolation and constituted the lower compartment limit, z_{pro}, for the mass balance (Fig. B5.1).

In the next step, PO_4^{3-} depletion between the surface layer and z_{pro} during the spring bloom was integrated over depth. The integral PO_4^{3-} depletion corresponding to the area between the red and blue lines in Fig. B5.1 provides a measure of the integrated net community production (iNCP), expressed as mol m^{-2}. Dividing the integrated PO_4^{3-} depletion by the PO_4^{3-} decrease in the upper surface yields a variable with the unit of length. This can be interpreted as the effective NCP penetration depth, z_{eff}, in the hypothetical case that the NCP in the surface layer is homogenously distributed over depth (green rectangle in Fig. B5.1). The integrated C_T depletion can then be obtained by multiplying the C_T depletion in the surface layer by z_{eff}, provided that PO_4^{3-} and C_T between the surface and z_{eff} are similarly affected by NCP. The latter assumption is not entirely fulfilled because C_T consumption during NCP is partly compensated for by gas exchange.

Daily air-sea fluxes, F_{AS}, were calculated according to Eq. 2.32 and integrated (summed) for the period between the start and end (May 25) of the spring bloom. Daily values for the necessary variables, i.e., SST, salinity, surface water pCO_2, and atmospheric pCO_2, were obtained by linear interpolation of the measured data. For pCO_2^{atm}, the 2005 measurements of CO_2 in dry air (Sect. 2.1) and an annual increase of 2 ppm (IPCC 2013) were used to obtain the CO_2 concentration in 2009. The pCO_2^{atm} was calculated by assuming water-vapour saturation at the sea surface and standard atmospheric pressure. Furthermore, mean daily wind speeds were calculated for areas embedding the considered sub-transects (Fig. 4.4) using the high-resolution wind fields (25 $\times$ 25 km) obtained from the coastdat databank, which provides re-analysis wind data for the Baltic Sea region (Weisse et al. 2015, http://www.coastdat.de). The net community production determined in this way refers to the production of both particulate and dissolved organic carbon (POC and

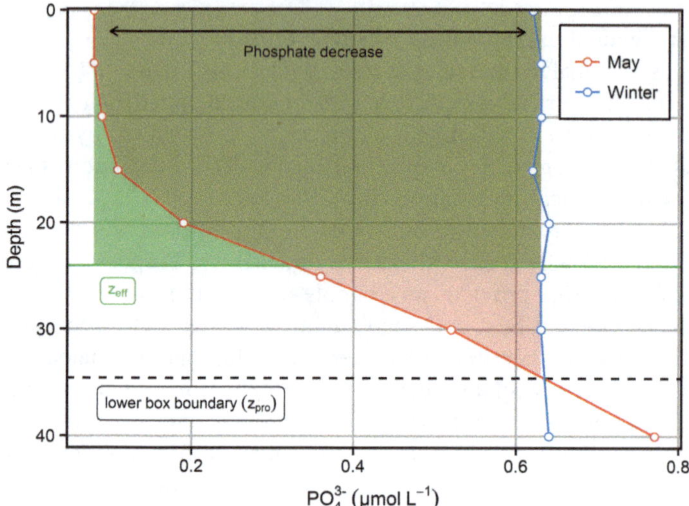

Fig. B5.1 Schematic presentation of the effective NCP penetration depth, zeff, derived from the [PO_4^{3-}] profile during winter (*blue line*) and after the spring bloom (*red line*).

DOC, respectively), $iNCP_t$, where production of POC is more important quantity because it controls the transport of carbon into deeper water layers and the associated biogeochemical transformations. In accordance with studies in the Gotland Sea (Schneider et al. 2003; Hansell and Carlson 1998), a value of ~20% was adopted for the amount of $iNCP_t$ transferred into the DOC pool and a factor of 0.8 applied to convert the results of Eq. 5.2 to the NCP of POC (Eq. 5.3).

$$iNCP = 0.8 \cdot iNCP_t \tag{5.3}$$

The integrated net community production of POC obtained in this way for the individual sub-transects are shown in Table 5.2 and Fig. 5.12a. Remarkable differences in the magnitude of the iNCP ranging between 840 mmol-C m^{-2} in the MEB and >2700 mmol-C m^{-2} for the EGS and HGF are observed. The important role played by nutrient concentrations in controlling iNCP intensity is discussed in the following section. However, another limiting factor for the iNCP is the sea's topography, in the case that the water depth is lower than that at which solar irradiation is still sufficient to allow net production. This partly explains the low iNCP for the MEB, where the mean water depth is ~20 m and thus confines primary production to this depth range. Another aspect for the assessment of the iNCP refers to the contribution of the CO_2 gas exchange term ($\int F_{AS}$, Eq. 5.2) to the mass balance. This is indicated in Fig. 5.11 (upper blue lines) where the sum of the observed C_T^* depletion and of the effect of the integrated (over time) gas exchange on C_T^* ($\int F_{AS}/z_{eff}$) are shown. The blue lines thus represent the C_T^* decrease in case that no CO_2 uptake by gas exchange would have occurred during the OM production. The contribution of the gas exchange term gains importance if

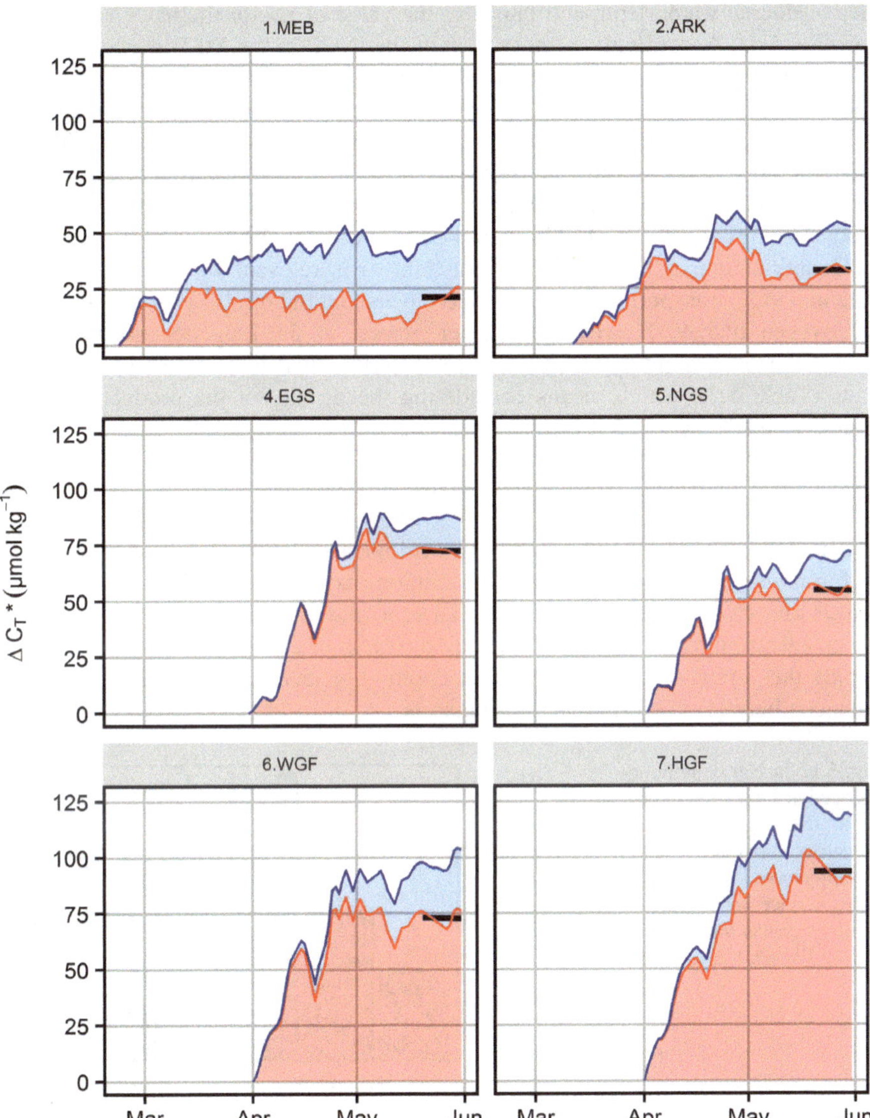

Fig. 5.11 Development of ΔC_T^* during the spring bloom in the different sub-transects (red lines). The mean ΔC_T^* at the end of May (black bar) is used for the production estimate (Eq. 5.2). Including the gas exchange term in the mass balance (expressed in concentration units, $1/z_{eff} \bullet \int F_{AS}\, dt$) - yields the upper blue line.

the considered water depth and thus also the value of z_{eff} for the $i\Delta C_T{}^*$ estimate is low (Eq. 5.2). Accordingly, almost 50% of the iNCP in the MEB is accounted for by the air-sea gas exchange term whereas ~20% is contributed in the more northern sub-transects. The explicit consideration of the air-sea gas exchange term is appropriate because calculations of CO_2 gas exchange are associated with considerable uncertainties. They are therefore the main source of errors in estimates of iNCP, in addition to the assumption concerning the 20% DOC contribution the total production (Schneider et al. 2003; Hansell and Carlson 1998).

Quantification of the iNCP (in mol-C m^{-2}) is relevant for an assessment of potential POC transport into deeper water layers and the resulting consequences for the oxygen budget. However, for other aspects of the biogeochemistry of OM production, it is more informative to characterize NCP in terms of concentration units (Table 5.2), which means considering the density of the productivity. It is related to iNCP by Eq. 5.4 (see also Eq. 5.2):

$$NCP = iNCP/z_{eff} \tag{5.4}$$

The latter influences the manifold ecological relationships and interactions within complex marine food web. In addition, expressing NCP in terms of concentrations is appropriate when its relation to nutrient availability is of interest. A consideration of the regional distribution pattern of the concentration-based NCP avoids the limitations posed by water depth (Fig. 5.12) and are therefore more balanced between the different sub-transects.

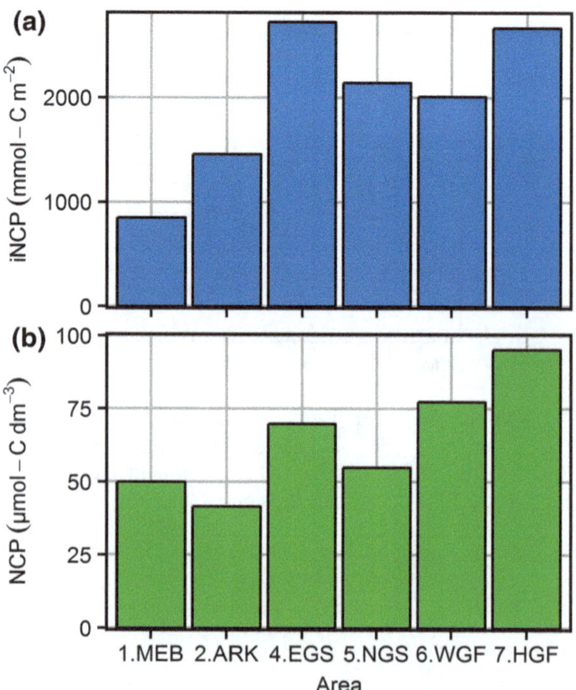

Fig. 5.12 **a** Net community production integrated over depth, iNCP, given as moles per surface area and **b** net community production in the upper surface layer, NCP, given as concentration units for the six sub-transects in 2009

5.2.3 Nitrogen Supply for the Spring Bloom

Nitrate and phosphate accumulate in the surface water during winter convective mixing. In the Baltic Proper, the ratio of the molar concentrations of N and P (N/P) is in the range of 5.5–8.0 and is thus 2- to 3-fold lower than the Redfield ratio (N/P = 16), i.e., the N/P demand of marine primary producers cells for optimal growth. Hence, the traditional view is that net OM production is nitrogen-limited; that is, it stops after the complete exhaustion of nitrate despite the continued availability of phosphate. However, this view conflicts with our observations of the CO_2 system.

Monthly measurements of the nitrate and phosphate concentrations in the central Gotland Sea (station BY15) are available through the Swedish Monitoring Programme (SMHI). The seasonal distribution of the nitrate concentrations in surface waters for the period 2000–2014 (Fig. 5.13a) showed a rapid decrease with the start of the spring bloom and an almost complete exhaustion by mid-April. In none of the years were the nitrate concentrations measured after April 15 large

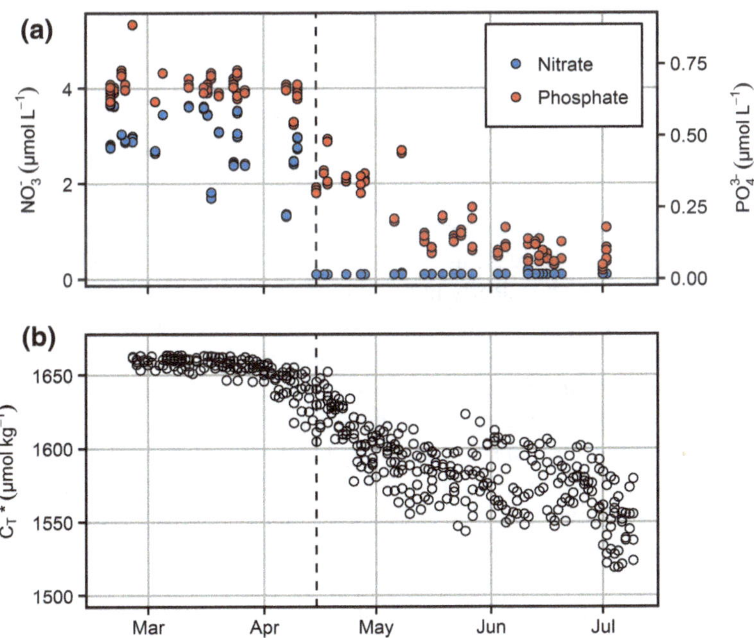

Fig. 5.13 **a** Surface water concentrations of nitrate and phosphate during the spring bloom period (monthly observations of the Swedish monitoring program (SMHI) at station BY-15 for the period 2003–2014). The *vertical line* indicates the date at which nitrate was completely exhausted throughout all years. In contrast, phosphorus is still available and successively consumed. **b** C_T* during the productive period in the Eastern Gotland Sea, merged for all years from 2003–2015

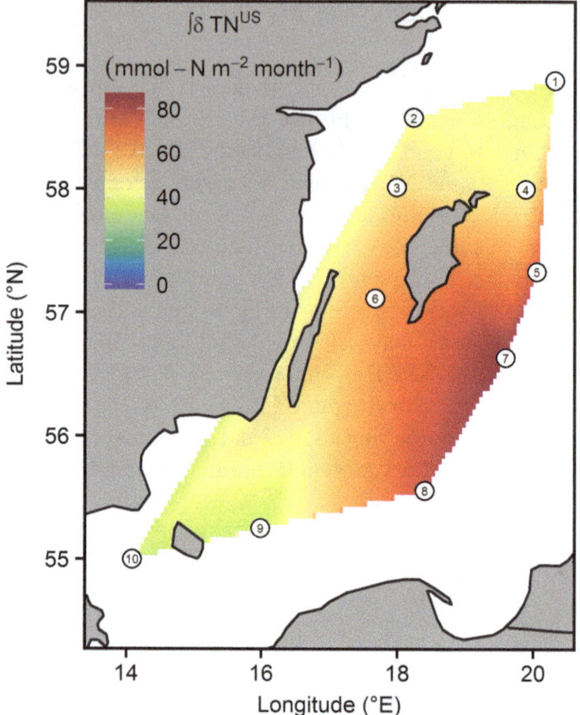

Fig. 5.14 Mean regional distribution of a nitrogen source which was identified for the spring bloom period in May ($NO_3^- = 0$) on the basis of a combined total N/total P mass balance (modified after Eggert and Schneider 2015). A nitrogen source could not be detected at stations in the Kattegat and the Gulf of Bothnia

enough to drive the further measurable net production of OM. However, in almost all years, the significant decrease in C_T* did not cease after April 15 (Fig. 5.13b) but continued during May. This clearly indicated continued net OM production and was consistent with a decrease in the phosphate concentration after nitrate depletion (Fig. 5.13a). This conclusion is also valid for the ΔC_T* during the spring of 2009, when automated sampling for the analysis of nutrient concentrations provided data with a high temporal resolution during the spring bloom period (Algaline Project, SYKE). Nitrate concentrations decreased to below 0.1–0.2 μmol L^{-1} already by mid-April in the four northern sub-transects, but ΔC_T* increased by a factor of ~2 after the nitrate depletion (Fig. 5.11). The exceptions were in the MEB and ARK, where increases in ΔC_T* after nitrate depletion could not be unambiguously detected.

Since the atmospheric deposition of inorganic nitrogen is by far too small to satisfy the nitrogen demand (Bartnicki et al. 2011) during the post-nitrate NCP, the question concerning the nitrogen supply was raised and has stimulated a controversial discussion within the Baltic Sea biogeochemical research community. The different hypotheses include the preferential mineralization of N contained in POM produced previously during the nitrate-based bloom. Alternatively, dissolved organic nitrogen (DON) compounds may provide nitrogen for new production, or protein-free organic matter may be produced, obviating the demand for nitrogen.

However, none of these processes convincingly and quantitatively explain the magnitude of post-nitrate production, which may even exceed production during the nitrate-fueled bloom.

A new aspect was introduced into the discussion when a budget for total nitrogen [N_T = DIN + dissolved organic N (DON) + particulate organic N (PON)] was established on the basis of data obtained from a central station in the EGS. Atmospheric DIN deposition (Bartnicki et al. 2011) and the loss, by sedimentation, of N_T derived from the loss of total phosphorus (P_T) indicated the existence of a late-spring external N source in the central Baltic Sea (Schneider et al. 2009). The latter could not be explained by vertical mixing because in late spring the thermocline is located at a depth of ~30 m whereas complete nitrate depletion occurs down to a depth of ~50 m. The lateral transport of nitrate from coastal or estuarine regions was excluded as a N source because horizontal nitrate gradients are virtually absent. Therefore, it was suggested that N-fixation takes place already in May. Since this contrasts with the well-documented N-fixation during mid-summer, when the SST is high, the earlier N-fixation has been termed "cold fixation." In a subsequent data analysis based on the same approach, Eggert and Schneider (2015) used the monthly N_T and P_T concentration data from the entire Swedish monitoring network in the Baltic Sea (11 stations) from 1995 to 2013 to confirm the previously identified N excess provided from an external source, such as N-fixation. A statistically significant N excess was detected in May and it extended over the entire station network in the Baltic Proper (Fig. 5.14), whereas there was no N excess in either the Kattegat or Gulf of Bothnia. The latter finding is consistent with the assumption that cold fixation can only fuel biomass production if phosphate is still available after nitrate depletion, which is not the case in the Kattegat and Gulf of Bothnia. The long-term mean of the May N source for the central Baltic Proper was 55 mmol m^{-2}, corresponding to ~40% of the mid-summer N source (141 mmol m^{-2}) that was no doubt caused by N-fixation (Wasmund et al. 2001).

In view of both the continued NCP after the exhaustion of nitrate and the existence of a N source, the assumption of cold fixation seems reasonable. However, the hypothesis proposing cold fixation has a serious flaw: Until now, neither species-level biological observations nor sporadic N-fixation-rate measurements based on the ^{15}N incubation method have provided evidence of cold fixation. Thus, the problem remains to be solved because post-nitrate production may be an important contribution to eutrophication of the Baltic Sea (see Sect. 5.2.2). The lack of understanding has also complicated efforts to simulate the spring-bloom decrease in pCO_2 by model calculations. Attempts to reproduce the continued pCO_2 decrease (C_T) after nitrate depletion have included fractional nutrient release (Omstedt et al. 2009), an internal redistribution of organic and inorganic N compounds (Kreus et al. 2015), and the introduction of an "unknown" N-fixing plankton species (Kuznetsov et al. 2011). However, there is as yet no observational support for any of these model modifications

5.2.4 *Net Community Production and Nutrient Consumption*

By measuring nitrate and phosphate concentrations during the spring bloom in 2009, it was possible to determine bulk C/N and C/P ratios for NCP and to compare them to the classical Redfield ratios (Fig. 5.15). The temporal resolution of the nutrient measurements was too low to resolve the gradual nitrate depletion during the nitrate-based bloom, which lasted ~2 weeks. The first available nutrient measurement after the start of the spring bloom already showed nitrate concentrations that were essentially zero. It was therefore assumed that the winter nitrate concentrations present at the start of the spring bloom had fueled the NCP determined at the date of the first zero-nitrate measurement. This is, of course, a simplification because the complete exhaustion of nitrate may have occurred prior to the measurements, when NCP was less advanced. Nonetheless, the plot of NCP from the different sub-transects versus the corresponding winter nitrate concentrations (Fig. 5.15b) showed a regression line with a slope only ~20% larger than that

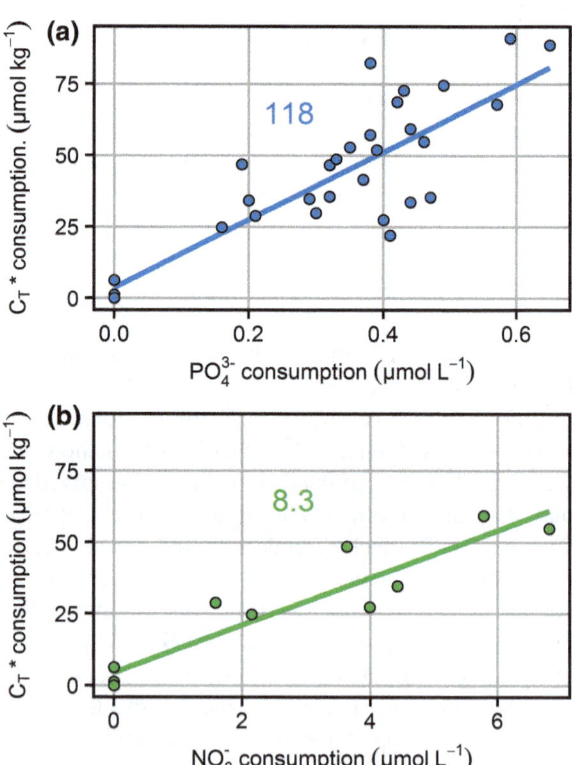

Fig. 5.15 C_T* consumption plotted as a function of **a** phosphate and **b** nitrate consumption during the 2009 spring bloom. Linear regression analysis revealed C/N and C/P ratios of 118 and 8.0, respectively

of the Redfield C/N ratio for marine POM (Redfield et al. 1963). However, this comparison should be interpreted with caution in view of the above-mentioned simplifications which may cause an overestimation of the C/N ratios. Furthermore, the nutrient data were obtained from only a few samples (2–5) within the individual sub-transects whereas the NCP was derived from the mean of continuous pCO_2 observations along the transect.

For the consumption of phosphate that occurred in accordance with the development of the spring NCP, PO_4^{3-} concentration data were available at five dates during the spring bloom and facilitated the estimation of the mean C/P ratio with respect to NCP and PO_4^{3-} consumption. Figure 5.15a presents NCP as a function of PO_4^{3-} consumption determined with reference to the winter PO_4^{3-} concentration; that is, the conditions at the start of the spring bloom. The slope of the regression line was 118 and exceeded the Redfield ratio (106) by ~10%, but was close to the C/P ratio (123) reported by Körtzinger et al. (2001). Despite the uncertainties associated with the comparison of the data from individual samples (PO_4^{3-}) with the mean of continuous records (NCP), our C/P ratio is within a biogeochemically reasonable range.

5.2.5 The "Blue Water" Period

The pCO_2 minimum during spring indicates the termination of the spring bloom NCP. In 2009, the subsequent increase in pCO_2 in the different sub-transects was evident already between the end of April and beginning of May (Fig. 5.5), whereas in other years the reversal of the pCO_2 trend occurred later in May. To estimate the role of seasonal warming on the temporary pCO_2 increase, which generally lasted until mid-June, we analyzed the relationship between $\ln(pCO_2)$ and SST in the different sub-transects. The slopes of the linear regression lines ranged from 0.06 to 0.11 K^{-1} for all sub-transects except the MEB, where a clear borderline between different production regimes was difficult to identify. In the example from the WGF given in Fig. 5.16, the slope of a linear regression line for the period of increasing pCO_2 was $d(\ln pCO_2)/dT = 0.073\ K^{-1}$. This value exceeded the theoretical temperature coefficient ($0.0423\ K^{-1}$, see Sect. 2.3.4) and indicated that, at constant A_T, increasing C_T* must have contributed to the increase in the pCO_2, which corresponded to the uptake of CO_2 from the atmosphere as evidenced by the stable level of the accumulated iNCP, shown for the second half of May in Fig. 5.11. The lack of NCP or its occurrence at a rate too low to be detected by our measurements does not rule out biological activity; rather, it indicates a balance between OM production and mineralization. Under these conditions, chlorophyll-a concentrations (Wasmund and Siegel 2008) are low and the color of the water is therefore more blue than green; hence the term "blue water" period.

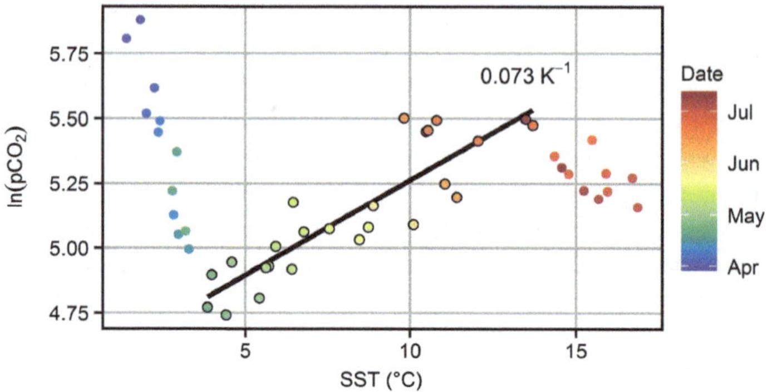

Fig. 5.16 Increase of the surface water pCO_2, presented as $\ln(pCO_2)$, during the 2009 "blue water" period in the western Gulf of Finland (WGF) as a function of the sea surface temperature. The preceding and following decrease of the pCO_2 are due to C_T* consumption during net community production

5.2.6 Control of Mid-Summer Net Community Production and N-fixation

In 2009, the first signs of a revival of NCP were observed by mid-June and production continued until the end of August. The C_T* for this period in the six sub-transects is shown in Fig. 5.17, together with SST development. Up to three distinct NCP pulses were indicated by a sudden drop of C_T* in the sub-transects between EGS and HGF and by the oxygen saturation data, which were available for the date of the C_T* minimum (July 2, 2009) in the EGS. The pCO_2 and O_2 saturation data along the EGS sub-transect were clearly inversely correlated (Fig. 5.18), with extreme O_2 saturation of almost 130% coinciding with a drop in the pCO_2 to <120 μatm. Unfortunately, a quantitative determination of whether the relationship between CO_2 depletion and O_2 production was consistent with the bulk stoichiometry of photosynthesis was not possible. This calculation must include the effect of CO_2 and O_2 gas exchange, which could not be estimated because of a lack of information on mixing of the surface layer (discussed in Sect. 5.2.7).

In the absence of DIN and in view of many biological studies and observations (Wasmund et al. 2001; Wasmund et al. 2005), the detected NCP events could be convincingly attributed to the effect of N-fixation in the sub-areas between the EGS and WGF. The situation was less clear in the ARK, because the suspected C_T* minima were small (Fig. 5.17) and not always distinguishable from the general variability caused by the influence of the different water masses. C_T* development in the MEB underwent a gradual slight decrease during June/July (Fig. 5.17), indicative of ongoing NCP but atypical for the occurrence of N-fixation as observed in the northern sub-transects. Furthermore, the attribution of NCP in the MEB to the effect of N-fixation is questionable because surrounding land-borne DIN sources may have fuelled the NCP. This does not imply the absence of N-fixation and

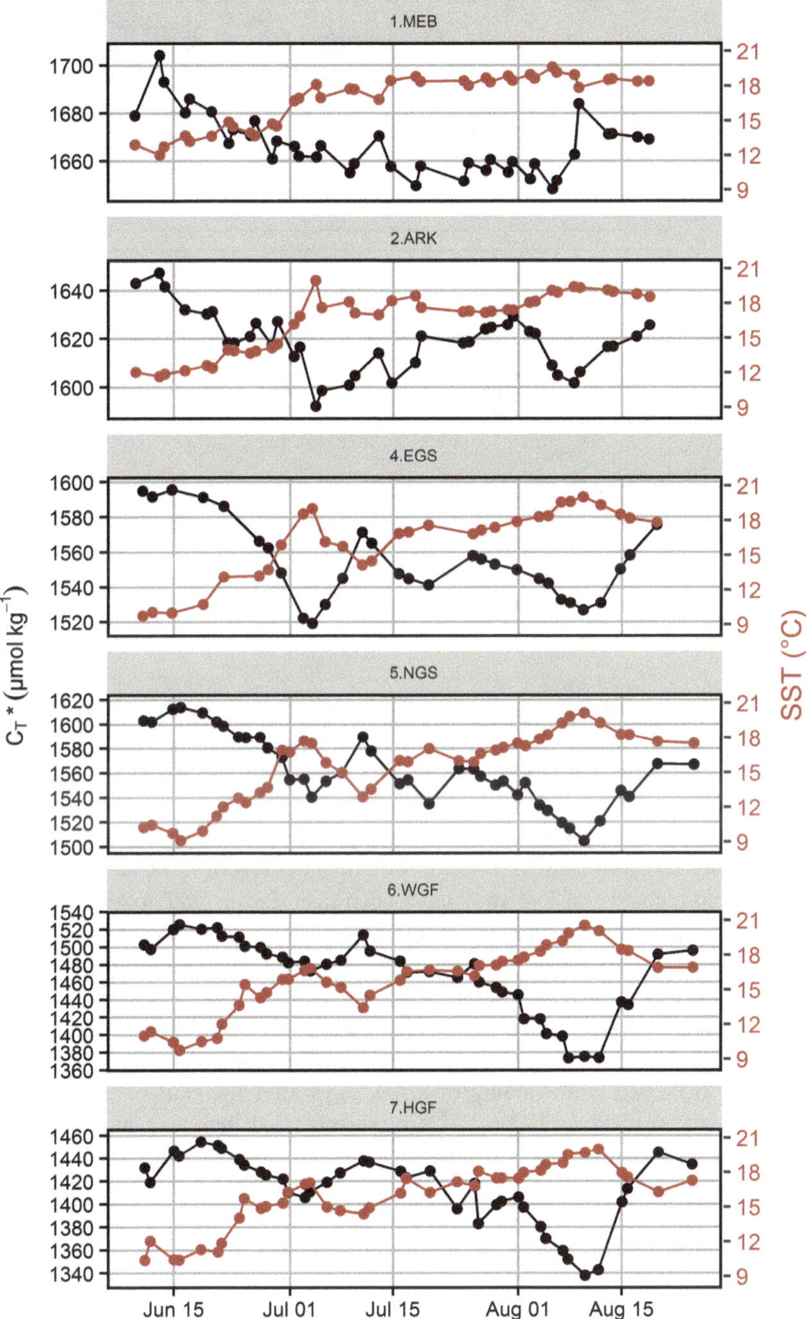

Fig. 5.17 Temporal changes in C_T* and sea surface temperature for the individual sub-transects during mid-June to the end of August 2009, a period during which production fuelled by N-fixation is common. The scaling of the C_T* axis is adjusted to the range of the data whereas the intervals of the *grid lines* are identical in all graphs in order to highlight the differences in the C_T* changes

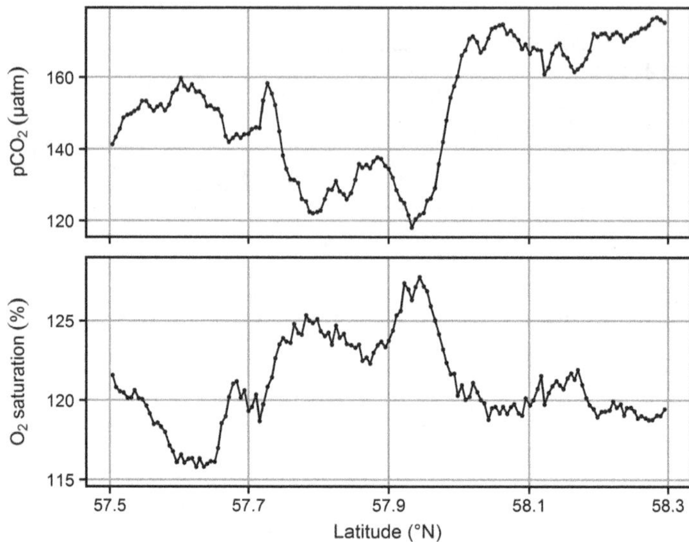

Fig. 5.18 Concurrent drop of the pCO_2 and increase of the O_2 saturation as consequence of a mid-summer production event fuelled by N-fixation in the Eastern (EGS) Gotland Sea in the beginning of July, 2009

related NCP in the MEB in general, as both the analysis of the pCO_2 data from other years and biological observations strongly indicated the existence of NCP supported by N-fixation in 2006 and 2008 (Schneider et al. 2015).

In the following, we focus on the individual C_T* depletion events that lasted for only 1–2 weeks and which in some cases (e.g., WGF, Fig. 5.17) caused a reduction of the C_T* of >100 µmol kg^{-1}, reflected in a mean pCO_2 of <100 µatm. However, the subsequent recovery of C_T* may have been even faster. SST followed the same cycles but in the opposite direction (Fig. 5.17). These patterns were in all cases related to the temporary low-wind conditions and thus to a change in wind-driven turbulence. Two examples that demonstrate the coincidence between the low C_T* and low wind speed are shown in Fig. 5.19 for NCP events that occurred in the NGS and WGF in early July and early August 2009, respectively. In both cases, the C_T* minima occurred with a delay of a few days after the start of a period when wind speeds were as low as 1–2 m s^{-1}. Conversely, C_T* increased again as soon as the wind speed exceeded 5–6 m s^{-1}. Similar relationships between wind speed and C_T* were determined in a previous pCO_2 data analysis that focused on the EGS in the years before 2011 (Schneider et al. 2014b). The authors concluded that N-fixation and the related NCP were strongly favored by a shallow surface layer that formed during calm weather conditions. The high surface NCP signal, as reflected in the lower C_T*, was diluted by increasing wind-driven vertical mixing, thus again increasing C_T*. This dependency of the NCP on mixing can be attributed to the control of N-fixation by solar radiation, which acts much more efficiently on the plankton particles residing in a shallow and stable surface layer and which are

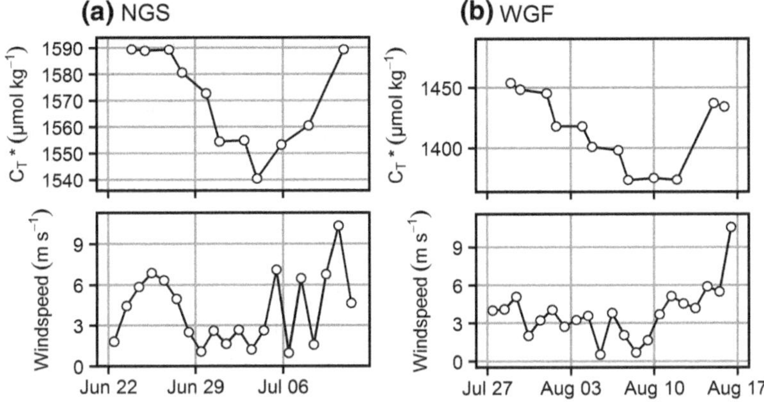

Fig. 5.19 Interaction between low wind speed and the development of mid-summer bloom events, illustrated by episodes observed for the sub-transect NGS (**a**) and WGF (**b**) in 2009

not subjected to vertical mixing. Likewise, the temperature increase, primarily driven by solar radiation, reacts upon changes in the mixing regime, which leads to the strong coupling of C_T* and SST during individual mid-summer NCP events. This relationship was previously analyzed on the basis of pCO_2 data from the EGS (Schneider et al. 2014b). C_T* depletion during several mid-summer production events and the concurrent uptake of atmospheric CO_2 were used to calculate NCP_t production, defined as the sum of NCP and the release of DOC. The latter was assumed to make up 20% of the total NCP_t production (Sect. 5.2.2). Distinct linear relationships between the accumulated NCP_t and SST were obtained for all considered years, including highly similar slopes ranging between 10 and 15 µmol kg^{-1} K^{-1} (Schneider et al. 2014b). Based on these relationships, it was shown that production rates were not controlled by temperature as such but by the rate of its increase, with both rates being controlled by the amount of solar radiation (Schneider et al. 2014b).

Here we present similar calculations for different sub-transects in 2009, except the MEB, where no clear mid-summer production events were recorded in that year. NCP_t for the individual C_T* depletion pulses was calculated according to Eq. 5.5:

$$NCP_t = \Delta C_T^* + \frac{1}{z_{mix}} \int_{t,start}^{t,end} F_{AS} \cdot dt \quad (5.5)$$

The air-sea gas exchange term was calculated as described above (see Sect. 5.2.2). Since no information was available concerning the vertical structure of the water column for the considered time periods, the depth of the sample-water inlet (3 m) on the cargo ship was used as a lower limit estimate for z_{mix} (see Sect. 5.2.6).

Accounting for a 20% DOC share, the NCP_t obtained by these calculations were consistent within the range of the POC concentrations observed at EGS in July/August (40–50 μmol-C dm^{-3}, IOW, Long-term observation programme) when taking into account the potential removal of POC by sinking. Furthermore, the plots of NCP_t as a function of SST for the most intense production pulses that were observed in the individual sub-transects, were again well-approximated by linear regression lines. The corresponding slopes for ARK, EGS, and NGS ranged between 10.4–14.4 μmol kg^{-1} K^{-1} (Fig. 5.20) and were similar to those determined previously for the EGS in different years. However, for the WGF and HGF the slopes were 26.3 and 29.8 μmol kg^{-1} K^{-1}, roughly twice as large as those in the more southern sub-transects. According to our interpretation of the coupling between NCP_t and SST, production and, hence, N-fixation in the WGF and HGF are more sensitive to the efficiency of solar radiation. However, this conclusion is speculative and not supported by any evidence.

Another important aspect of the previous analysis of mid-summer production in the EGS is related to the factors that limit mid-summer production driven by N-fixation. For all production events, C_T* decreased as long as the SST increased (Schneider et al. 2014b). The findings for mid-summer production in the different sub-transects in 2009 were similar (Fig. 5.17) except for the MEB, which lacked evidence of distinct production events. These observations challenge the traditional view that the amount of excess phosphate persisting after the spring bloom is the limiting factor for N fixation and related production. The minor role of a PO_4^{3-} excess in mid-summer N-fixation is further supported by the fact that in many years the bulk of the PO_4^{3-} excess was already consumed in May (Swedish National Monitoring Programme, SMHI). For example, nutrient measurements performed on June 10, 2009 in conjunction with the pCO_2 records from the cargo ship showed a PO_4^{3-} concentration below the detection limit in all considered sub-transects. Therefore it was speculated that an input of PO_4^{3-} into the surface water had occurred by vertical mixing, upwelling, or by release from an existing pool of organic phosphorus (Nausch et al. 2009). However, assuming that PO_4^{3-} uptake followed the Redfield ratio (C/P = 106), then none of these potential P sources explains the observed mid-summer NCP (Schneider et al. 2003). The absence of an external PO_4^{3-} source that could satisfy a Redfield-like P production demand is also reflected in the vertical distribution of N_T and P_T (Fig. 5.21) measured within the SMHI monitoring program on the same day (July 2, 2009) when our pCO_2 (C_T*) measurements indicated high N-fixation activity. The latter was reflected in N_T increases of 6.3 μmol L^{-1} and 4.4 μmol L^{-1} at the immediate surface and a depth of 5 m, respectively, compared to the uniform background level below 5 m. At the same time, the P_T was only 0.07 μmol L^{-1} and 0.05 μmol L^{-1}, respectively, above the background concentration. These differences are by about one order of magnitude below the values that would be required for a Redfield-like production (Fig. 5.21).

On the other hand, several investigations have reported C/P ratios for the POM produced during N-fixation of ~500, far higher than the Redfield ratio (Larsson et al. 2001; Nausch et al. 2008). Hence, the search for a Redfield-conforming P

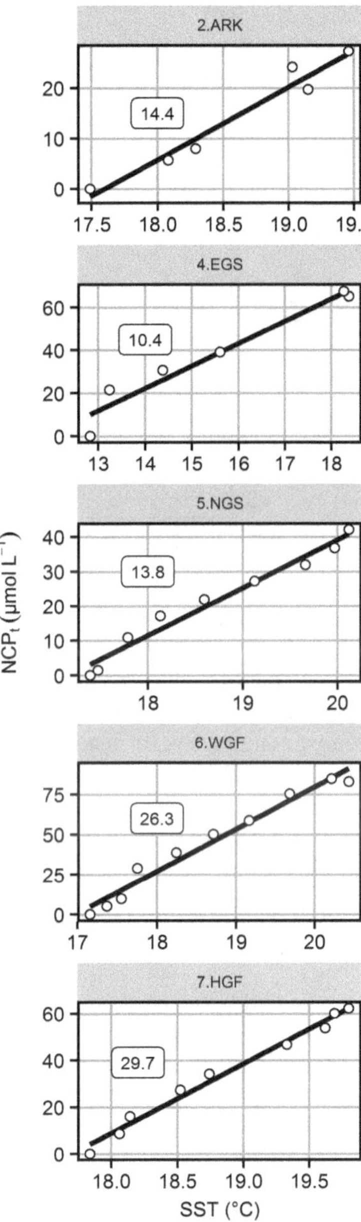

Fig. 5.20 Relationship between the accumulated net community production and the sea surface temperature for the main mid-summer production events in the individual sub-transects including linear regression lines. In the Mecklenburg Bight (MEB) distinct production pulses could not be detected

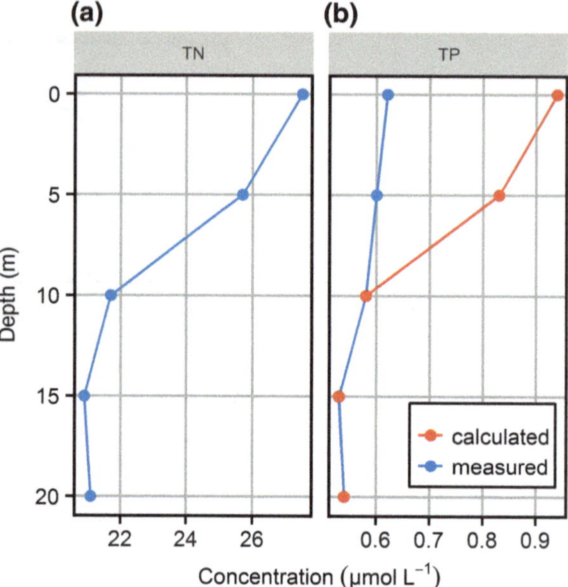

Fig. 5.21 Vertical profiles of **a** total nitrogen and **b** total phosphorus at station BY15 in the eastern Gotland Sea on July 2 (2009). The increase of total nitrogen in the surface caused by N-fixation was not accompanied by the input of an equivalent total P (PO_4^{3-}) according to the Redfield C/P ratio (**b**, *red curve*). This questions the hypothesis that nitrogen fixation is fueled by the input of PO_4^{3-} from deeper water layers

source is misleading and the role of phosphorus in production fueled by N-fixation remains unclear. A related question that must be addressed is the nature of the OM produced during N-fixation, as the C/P ratios of the POM do not follow any comprehensible production stoichiometry; instead, the Redfield-like C/N ratios (7.5) are conservative (Schneider et al. 2003). While this has been explained by the continued production of proteins in the absence of PO_4^{3-} (Schneider et al. 2014b), this conclusion remains to be verified in further investigations.

5.2.7 Estimation of Depth-Integrated N-fixation

Integrated N-fixation was quantified based on the related iNCP, which is somewhat more uncertain than the corresponding calculations for the spring bloom. The uncertainty arises from a lack of data, such as phosphate levels for the spring bloom, that indicate the penetration depth of NCP. In an earlier analysis of the pCO_2 data with respect to N-fixation, the depth of the mid-summer "historical" thermocline (Schneider et al. 2009) or a modeled penetration depth of surface processes (Schneider et al. 2014b) was used to determine the iNCP. The approach applied herein accounts for the occurrence of individual production pulses. As pointed out before, C_T* depletion must be confined to a shallow surface layer because mixing triggered by slight increases of the wind speed (Fig. 5.19) caused a reversal of the C_T* trend and resulted in a rapid increase of C_T*. To estimate NCP, we assumed that production took place only during C_T* depletion, because the radiation conditions during the subsequent mixing did not allow the continuation of N-fixation. The next challenge was to estimate the depth distribution of the C_T*

depletion, ΔC_T*. All that was known was that ΔC_T* refers to the water entering the ship at the depth of the inlet, which was ~3 m; thus, this value was used as z_{mix} to calculate iNCP whereas NCP below a depth of 3 m was ignored, as was the fact that the upper 3 m of the water column were not perfectly mixed. These shortcomings, together with the previous assumption that no NCP occurs during mixing after a production event, imply that the estimates of iNCP and the integrated N fixation (iNfix) constitute the lower limits. Accordingly, iNCP was calculated separately for each production event during the period from mid-June until the end of August, according to Eq. 5.6:

$$iNCP_t = z_{mix} \cdot \Delta C_T^* + \int_{t,start}^{t,end} F_{AS} \cdot dt \tag{5.6}$$

where the air-sea gas exchange term is calculated just as for the spring bloom period (Sect. 5.2.2). Annual mid-summer $iNCP_t$ was then obtained from the sum of the individual production events. Multiplication by a factor of 0.8 yielded the particulate fraction of (NCP), presented in Table 5.3 together with the sum of the spring and mid-summer NCP. The latter was in the range of 2.7–3.4 mol m^{-2} for the northern sub-transects, with a relatively uniform contribution by the N-fixation fuelled iNCP of 20–26%.

To calculate the N demand for the mid-summer production, iNCP was divided by the C/N ratio of POM (7.5) determined during a N-fixation period in the EGS (Schneider et al. 2003) and confirmed by long-term monitoring data (IOW). The corresponding data represent the annual mid-summer N-fixation integrated over depth (Table 5.3), iNfix. Except in the ARK, the values of iNfix in 2009 were surprisingly uniform, ~90 mmol m^{-2}. The value for the ARK was lower, consistent with biological observations suggesting decreasing N fixation towards higher salinities (Wasmund 1997). According to several previous estimates based on pCO_2 data from different years and focusing on the EGS, iNfix was roughly

Table 5.3 Comparison between the total (spring + mid-summer) integrated net community production (iNCP) and the iNCP during the mid-summer period, which is used to estimate the mid-summer nitrogen fixation iNfix. The data refer to 2009

sub-transect	iNCP, total (mmol-C m^{-2})	iNCP, mid-summer (mmol-C m^{-2})	iNCP, mid-summer (%)	iNfix (mmol-N m^{-2})
MEB	837	–	–	–
ARK	1867	453	24	60
EGS	3396	687	20	92
NGS	2830	679	24	91
WGF	2693	700	26	93
HGF	3395	698	21	93

between 100 and 200 mmol-N m^{-2} (Schneider et al. 2014b). This value was consistent with the N-fixation rates measured using the ^{15}N tracer method (138 mmol-N m^{-2}; Wasmund et al. 2005). The wide-ranging iNfix values can partly be attributed to interannual variability but were certainly also due to differences in data handling in the above-mentioned studies. This becomes obvious in a comparison of iNfix estimates for the EGS in 2009: 99 mmol-N m^{-2} based on the data analysis presented herein versus 214 mmol-N m^{-2} in a previous study (Schneider et al. 2014b). This discrepancy is the product of different assumptions concerning the vertical spreading of N-fixation. While we used the depth of the water inlet to obtain a lower limit estimate for iNfix, the model-based approach (Schneider et al. 2014b) produced a much larger penetration depth for C_T* depletion by N-fixation. This larger value may reflect the fact that the considered processes take place over a depth range of only a few meters and are difficult to resolve by models.

5.2.8 Autumn Mixing and Upwelling: The Occurrence of a Last Bloom Event

During the transition from late summer to autumn, the heat balance of the surface water becomes negative, such that SST decreases and thermal stratification becomes less stable. Together with an increasing probability of higher wind speeds, this causes an erosion of the summer thermocline and a progressive vertical mixing that is finally stopped at the permanent halocline in February/March of the following year. The start of the autumn mixing coincides with termination of the productive period. Since the water below the thermocline (the intermediate water) is highly enriched in C_T, due to OM mineralization and exchange across the halocline, vertical mixing causes a strong increase in C_T and thus in the pCO_2. This is

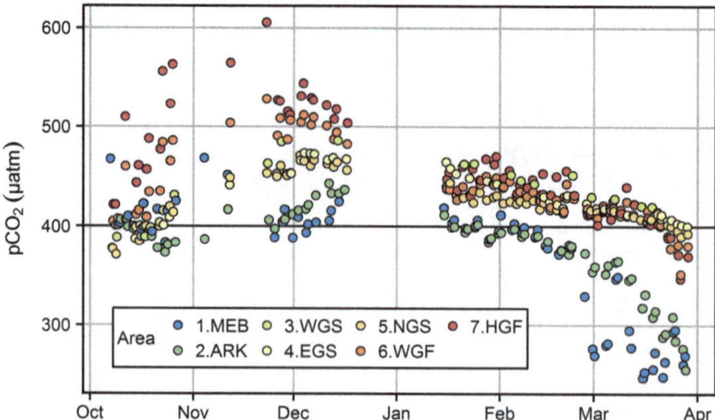

Fig. 5.22 The pCO_2 development in the different sub-transects between the onset of the autumn/winter vertical mixing in 2008 and the start of the spring bloom in 2009. The black line shows the atmospheric pCO_2. The data gap at the turn of the year is caused by technical reasons

illustrated in Fig. 5.22, which shows the pCO_2 for the period from October 2008 to March 2009 for all sub-transects, albeit with a larger data gap at the turn of the year. The increase of the pCO_2 differed considerably in the different sub-transects, with values only slightly exceeding atmospheric levels in MEB and ARK but reaching peak values of ~600 µatm in HGF. These differences are due to the extent of OM mineralization taking place in water layers subject to mixing and to the diffusion of C_T into these layers from below the permanent halocline. The import of C_T into the surface layer therefore depends on the on-site NCP but is also affected by the enrichment of C_T below the halocline, which is highly influenced by the lateral bottom transport of POM into the deep basins.

Another characteristic of the autumn pCO_2 increase was the large temporal variability during the first phase of vertical mixing. It was most pronounced in the HGF, where lateral transport plays a major role because of its proximity to the coastal heterogeneities caused by direct contact between the water column and the sediment surface, or by upwelling events. Alternatively, there may have been an autumn plankton bloom event that intermittently stopped the pCO_2 increase related to vertical mixing (see below). However, both the temporal variability of the pCO_2 and the regional differences decreased with time. Moreover, the pCO_2 values measured in the EGS, NGS, WGF, and HGF ultimately converged, reaching approximately the atmospheric level by the end of March, shortly before the start of the spring bloom. This behaviour was clearly due to CO_2 gas exchange and cooling, which levelled out the initial differences between sub-transects by generating an equilibrium with the atmosphere. The temporal pCO_2 pattern was less clear in the MEB and ARK, where atmospheric pCO_2 levels occurred a few weeks earlier because the lower water depth accelerates equilibration with atmospheric pCO_2.

In addition to convective and wind-driven mixing, water is transported from deeper layers by upwelling to the surface, but in general not from below the permanent halocline (Leppäranta and Myrberg 2009). Upwelling in the Baltic Sea takes place mostly in coastal regions, it is mainly wind-driven and takes place throughout the year, but it is more frequent during autumn and winter, when higher wind speeds are more common (Myrberg and Andrejev 2003). However, the effect on the composition of the surface water is most pronounced in early autumn, when a thermocline is still present and has separated the surface water from the intermediate water. The latter is highly enriched in the products of OM mineralization, such as C_T and nutrients. Enormous local increases in the pCO_2 of the surface water are occasionally observed and coincide with extremly low oxygen saturation of the surface water. This was the case on October 14, 2009, when heavy winds from the northeast (16 m s^{-1}) caused a strong upwelling off the northeast coast of the Island of Gotland. The pCO_2 rose to almost 2500 µatm within a narrow band of about 10 nautical miles (Fig. 5.23). This was higher than in any region of the Baltic Sea during the 12 years in which pCO_2 data were recorded by the cargo ship, as was also the case for O_2 saturation, which was as low as ~60% (Fig. 5.23). In accordance with the biogeochemical signals, the upwelling caused a drop in the SST of 4–6 K (Fig. 5.23).

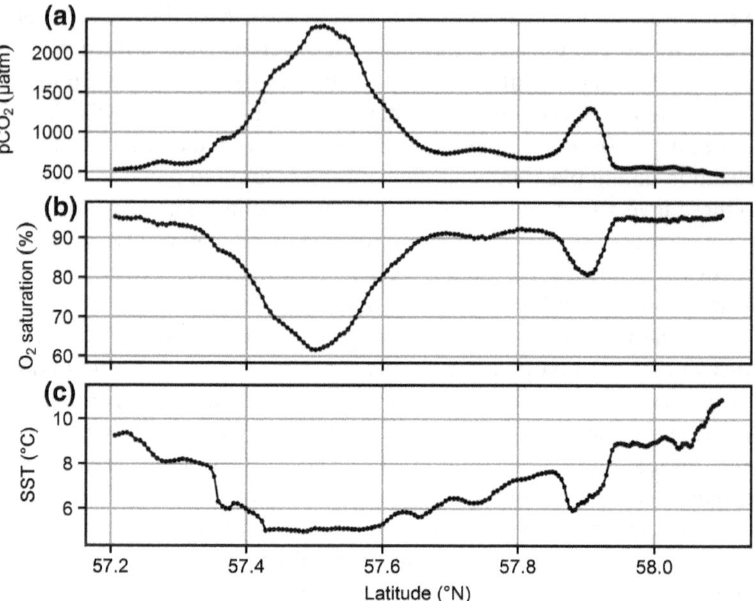

Fig. 5.23 An extreme upwelling event observed in October 2009 at the northwestern tip of the Island of Gotland resulted in **a** unusual high pCO_2 of almost 2500 µatm and **b** O_2 saturation as low as about 60%. At the same time **c** the sea surface temperature dropped by 4–6 K

The biogeochemical importance of upwelling events for the Baltic Sea as a whole is unclear and cannot be resolved by our data. Upwelling may simply constitute an early vertical mixing and thus contribute only to biogeochemical regeneration of the surface water, otherwise driven by basin-wide autumn/winter mixing. But it may also trigger local bloom events, which are relevant for biogeochemical cycles in the Baltic Sea.

Production events that do not follow the conventional spring/summer succession may also be triggered by convective mixing during autumn. These "autumn blooms" occur if the mixing is deep enough to transport nutrients to the surface, but does not exceed the critical depth (see 5.2.1), and happens during a time of year when the daily solar radiation is still sufficient to allow NCP. We have occasionally observed autumn blooms in the MEB, where their occurrence is favored by the early input of nutrients to the surface due to the low water depth. Figure 5.24 shows the pCO_2 along the MEB sub-transect for three consecutive cruises. For some sections of the transect of October 8, 2009, pCO_2 was clearly below the atmospheric level of ~400 µatm. The only plausible explanation is CO_2 consumption by an autumn bloom, as also wind speeds were low during the suspected production events. However, because the autumn bloom was detected only as a local and short-term event, quantitative estimates were not possible.

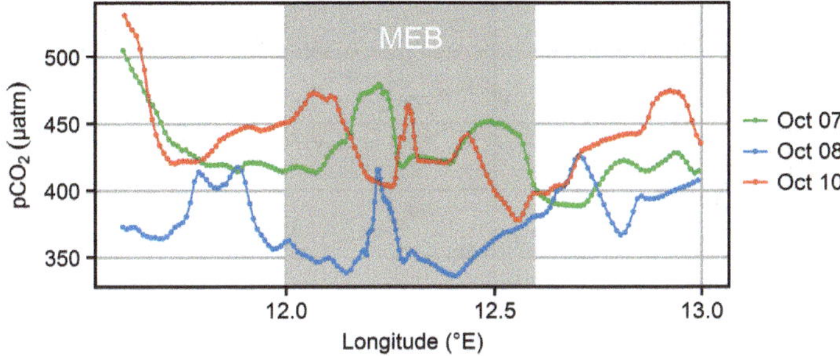

Fig. 5.24 The pCO_2 along three consecutive transects in October 2009 in the Mecklenburg Bight. On October 8 (*blue line*), the pCO2 (350 µatm) dropped several times considerably below the atmospheric pCO_2 of about 400 µatm. This is possibly the consequence of an autumn bloom event

5.2.9 Annual C_T* Cycling Presented as C_T* Versus SST Diagrams

The seasonality of C_T* in the different sub-transects in 2009 is presented in Fig. 5.5, where C_T* is plotted as a function of the date. In the following, we propose an alternative presentation of the seasonal cycle of C_T*, as C_T* versus SST diagrams (Fig. 5.25). The graphs indeed have the characteristics of a cycle because the C_T* at the starting date in mid-January and at the end of the year are almost identical in all sub-transects. This implies a steady state for the marine CO_2 system on an annual basis and that the effect of increasing atmospheric CO_2 on total CO_2 is not detectable within a single year. A steady state could also be assumed for other biogeochemical variables, such as nutrient concentrations, if trends in the inputs are ignored. Still, nutrient concentrations may differ considerably from year to year because of the potentially large interannual variability of nutrient inputs and internal transformation. Hence, a steady state becomes manifest only if the respective concentrations are averaged over many years. The CO_2 system differs because the same status is consistently reached in late winter of every year. This is a consequence of the moderating role of the air-sea gas exchange of CO_2, which reacts upon any interannual variability in the CO_2 system by pushing it towards equilibrium with the atmosphere.

The horizontal stretching of the graphs (Fig. 5.25) corresponds to the seasonal SST changes, which in the different sub-transects did not greatly differ in 2009 and ranged between 0 and 20 °C. This is in contrast to the vertical dimensions of the "cycles" that reflect seasonal C_T* amplitudes, which are controlled by different processes, as discussed above. Hence, our "walk through the seasons" can also be undertaken and summarized by following the C_T* versus SST pattern.

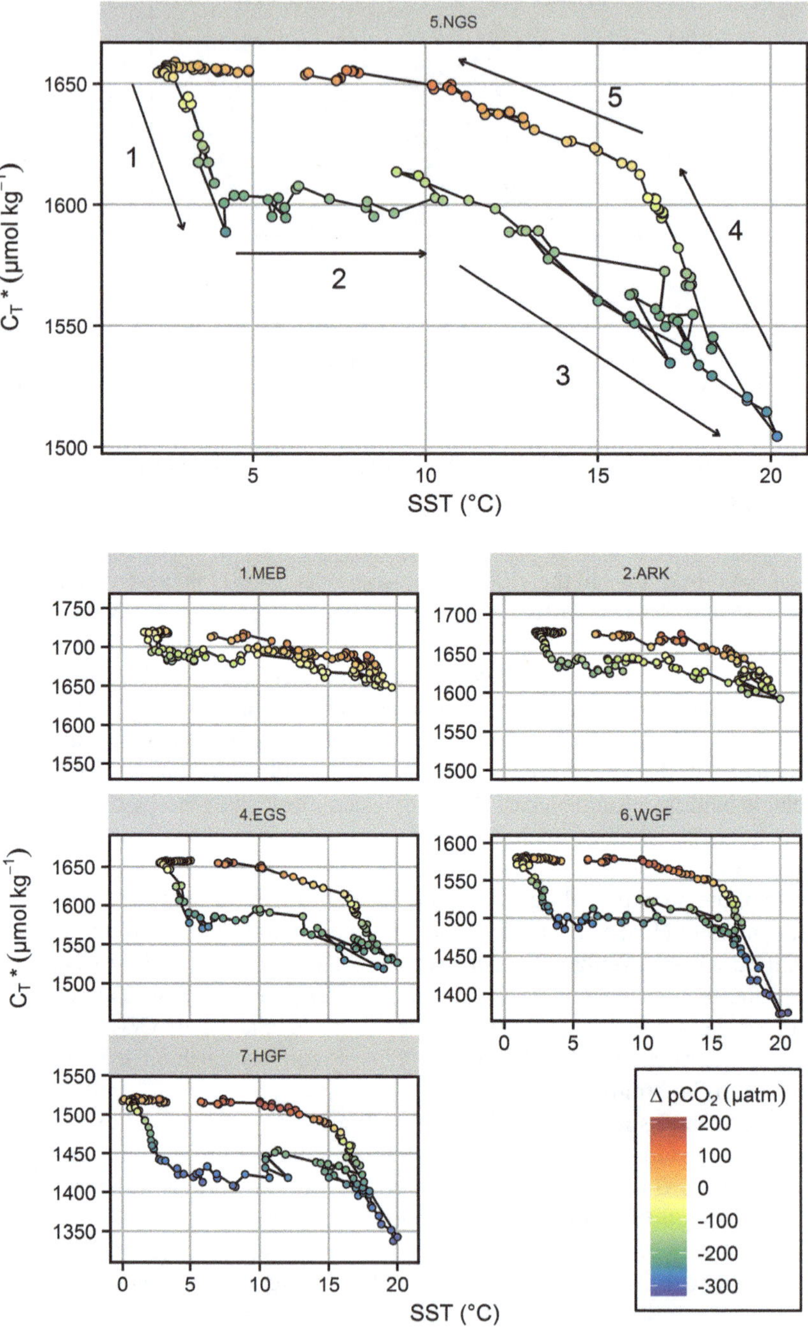

Fig. 5.25 Regional characterization of the seasonal cycle of the CO_2 system by C_T*—SST diagrams. For NGS different phases of the C_T* cycle are indicated: *1* spring bloom, *2* blue-water period, *3* midsummer NCP (N-fixation), *4* erosion of the thermocline, *5* further vertical mixing and CO_2 release by gas exchange. The color scale represents the CO_2 partial pressure difference between water and atmosphere, ΔpCO_2

Since the C_T*-SST diagrams for regions north of the ARK showed the clearest patterns, one such region, the NGS, serves as an example in the following. Tracking the C_T*-SST line from the starting point on January 15 shows that SST initially decreased by ~3 °C while C_T* remained almost constant during the same late winter period. However, with the reversal of SST development, and thus the start of the spring bloom, CT* decreased and a distinct linear relationship between C_T* and SST emerged. This relationship provides evidence for the important role of solar radiation in the control of NCP during the spring bloom. The entire spring bloom lasted over a SST range of only 2 °C, before the "blue water" period started at temperatures of 4–5 °C. The first signs of mid-summer NCP appeared already at 10 °C and marked the end of the "blue water" period. The following decrease in C_T* was characterized by several short-term pulses in which an increase in the SST coincided with a decrease in C_T*, together indicating the typical mid-summer N-fixation.

The subsequent increase in C_T* during vertical mixing and cooling occurred rapidly in the initial phase but then slowed down when the SST dropped below ~15 °C. This can be explained by two different cooling processes. In the beginning, wind-driven mixing and erosion of the thermocline may have controlled the decrease in the SST because extreme temperature gradients ≥ 1 K m^{-1} are common across the mid-summer thermocline. However, autumn/winter cooling of the surface water is also a consequence of the negative energy balance, an effect that gains increasing importance over the course of progressing mixing. But cooling as such does not reverse NCP, such that the increase in C_T* slows down during the the decrease in SST. The change occurs when SST reaches ~15 °C. Close to this temperature the CO_2 saturation state changes as well, from undersaturation to oversaturation and CO_2 uptake is replaced by CO_2 release. Therefore, the decrease in the C_T*-SST slope at 15 °C is reinforced by CO_2 gas exchange, which finally brings the surface water back to equilibrium with the atmosphere and closes the seasonal C_T*-SST cycle.

The different phases in the C_T*-SST diagram are presented here for the NGS but, in principle, characterize the other considered sub-transects as well (Fig. 5.25). However, there are also distinct differences between the individual sub-transects, especially with respect to the intensity of NCP during the spring bloom and the mid-summer N-fixation period. The C_T*-SST diagrams derived from high-resolution measurements of the surface water pCO_2 are therefore a useful tool to characterize regional differences in the NCP. They become even more meaningful if additional information on the vertical physical and biogeochemical structure of the water column is available. However, C_T*-SST diagrams may also be used to visualize the interannual variability and, perhaps, long-term trends in surface-water productivity including the relationship between temperature and productivity. For example, in 2013 and 2014 huge differences in the mid-summer NCP fuelled by N-fixation occurred in the NGS (Fig. 5.26) and in the other

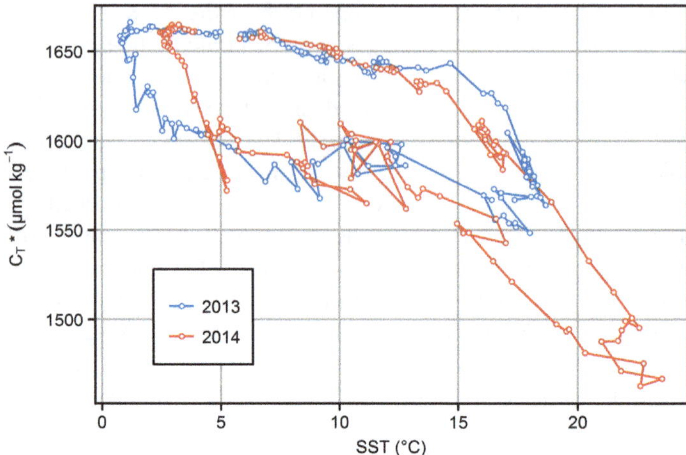

Fig. 5.26 Illustration of the interannual variability of the CO_2 system by the use of C_T*—SST diagrams, two examples from NGS in 2013 and 2014

sub-transects. In summary, high-resolution pCO_2 measurements are not only useful for detailed analyses of biogeochemical processes but also constitute a powerful monitoring tool for the assessment and illustration of trends in biological productivity and eutrophication.

References

Bartnicki J, Semeena V, Fagerli H (2011) Atmospheric deposition of nitrogen to the Baltic Sea in the period 1995–2006. Atmos Chem Phys 11:10057–10069

Bates NR, Best MHP, Neely K, Garley R, Dickson AG, Johnson RJ (2012) Detecting anthropogenic carbon dioxide uptake and ocean acidification in the North Atlantic Ocean. Biogeosciences 9:2509–2522

Eggert A, Schneider B (2015) A nitrogen source in spring in the surface mixed-layer of the Baltic Sea: evidence from total nitrogen and total phosphorus data. J Mar Sys 148:39

Gustafsson E, Wällstedt T, Humborg C, Mörth M, Gustafsson BG (2014) External total alkalinity loads versus internal generation: the influence of nonriverine alkalinity sources in the Baltic Sea. Glob Biogeochem Cycles 28:1358–1370

Hansell DA, Carlson CA (1998) Net community production of dissolved organic carbon. Glob Biogeochem Cycles 12(3):443–453

IPCC, 2013. Summary for policymakers, p. 27. In: Stocker TF, Qin D, Plattner GK, et al. (eds) Climate change 2013: the physical science basis. Contribution of working group I to the fifth assessment report of the Intergovernmental Panel on Climate change. Cambridge University Press

Kreus M, Schartau M, Engel A, Nausch M, Voss M (2015) Variations in the elemental ratio of organic matter in the central Baltic Sea: part I—linking primary production to remineralization. Cont Shelf Res 100:25–45

Körtzinger A, Hedges JI, Quay PD (2001) Redfield ratios revisited: removing the biasing effect of anthropogenic CO_2. Limnol Oceanogr 46(4):964–970

Kuznetsov I, Neumann T, Schneider B, Yakushev E (2011) Processes regulating pCO_2 in the surface waters of the central eastern Gotland Sea: a model study. Oceanol 53:745–770

Larsson U, Hajdu S, Walve J, Elmgren R (2001) Baltic Sea nitrogen fixation estimated from the summer increase in the upper mixed layer total nitrogen. Limnol Oceanogr 46:811–820

Leppäranta M, Myrberg K (2009) Physical oceanography of the Baltic Sea. Springer-Verlag, Berlin

Millero FJ (2010) Carbonate constants for estuarine waters. Mar Freshwater Res 61:139–142

Myrberg K, Andrejev O (2003) Main upwelling regions in the Baltic Sea—a statistical analysis based on three-dimensional modelling. Boreal Environ Res 8:97–112

Nausch M, Nausch G, Lass H-U, Mohrholz V, Nagel K, Siegel H, Wasmund N (2009) Phosphorus input by upwelling in the eastern Gotland Basin (Baltic Sea) Gotland basinin summer and its effects on filamentous cyanobacteria. Estuar Coast Shelf Sci 83:434–442

Nausch M, Nausch G, Wasmund N, Nagel K (2008) Phosphorus pool variations and their relation to cyanobacteria development in the Baltic Sea: a three-year study. J Mar Sys 71:99–111

Omstedt A, Gustafsson E, Wesslander K (2009) Modelling the uptake and release of carbon dioxide in the Baltic Sea surface water. Cont Shelf Res 29:870–885

Platt T, Lewis M, Geider R (1984) Thermodynamics of the pelagic ecosystem: elementary closure conditions for biological production in the open ocean. In: Fasham MJR (ed) Flows of energy and materials in marine ecosystems. Plenum Press, New York, pp 49–84

Redfield, AC, Ketchum, BH, Richards, FA (1963) The influence of organisms on the composition of sea water. In: Hill MN (Ed) The Sea Interscience, vol 2, New York, pp 26–77

Schneider B, Nausch G, Nagel K, Wasmund N (2003) The surface water CO_2 budget for the Baltic Proper: a new way to determine nitrogen fixation. J Mar Sys 42:53–64

Schneider B, Kaitala S, Maunula P (2006) Identification and quantification of plankton bloom events in the Baltic Sea by continuous pCO_2 and chlorophyll a measurements. J Mar Sys 59:238–248

Schneider B, Kaitala S, Raateoja M, Sadkowiak B (2009) A nitrogen fixation estimate for the Baltic Sea based on continuous pCO_2 measurements on a cargo ship and total nitrogen data. Cont Shelf Res 29:1535–1540

Schneider B, Nausch G, Pohl C (2010) Mineralization of organic matter and nitrogen transformations in the Gotland Sea deep water. Mar Chem 119:153–161

Schneider B, Gülzow W, Sadkowiak B, Rehder G (2014a) High potential of VOS-based measurements in Baltic Sea surface waters for detecting sinks and sources of carbon dioxide and methane. J Mar Sys 140:13–25

Schneider B, Gustafsson E, Sadkowiak B (2014b) Control of the mid-summer net community productionNet community production and nitrogen fixation in the central Baltic Sea: an approach based on pCO2 measurements on a cargo ship. J Mar Sys 136:1–9

Schneider B, Buecker S, Kaitala S, Maunula P, Wasmund N (2015) Characteristics of the spring/summer production in the Mecklenburg Bight (Baltic Sea) as revealed by long-term pCO2 data. Oceanologia 57:375–385

Sverdrup HU (1953) On conditions for the vernal blooming of phytoplankton. J Cons Int Explor Mer 18:287–295

Tyrrell T, Schneider B, Charalampopoulou A, Riebesell U (2008) Coccolithophores and calcite saturation state in the Baltic and Black Seas. Biogeosciences 5:1–10

Wasmund N (1997) Occurrence of cyanobacterial blooms in the Baltic Sea in relation to environmental conditions. Int Revue ges Hydrobiol 82:169–184

Wasmund N, Nausch G, Matthäus W (1998) Phytoplankton spring blooms in the southern Baltic Sea—spatio-temporal development and long-term trends. J Plankton Res 20:1099–1117

Wasmund N, Voss M, Lochte K (2001) Evidence of nitrogen fixation by non-heterocystous cyanobacteria in the Baltic Sea and re-calculation of a budget of nitrogen fixation. Mar Ecol Prog Ser 214:1–14

Wasmund N, Nausch G, Schneider B, Nagel K, Voss M (2005) Comparison of nitrogen fixation rates determined with different methods: a study in the Baltic Proper. Mar Ecol Prog Ser 297:23–31

Wasmund N, Siegel H, 2008. Chapter 15, Phytoplankton. In Feistel R, Nausch G, Wasmund N (eds) State and Evolution of the Baltic Sea, 1952–2005. A detailed 50-year survey of meteorology and climate, physics, chemistry, biology, and marine environment, Wiley, pp 441–481

Weisse R, Bisling P, Gaslikova L, Geyer B, Groll N, Hortamani M, Matthias V, Maneke M, Meinke I, Meyer E, Schwichtenberg F, Stempinski F, Wiese F, Wöckner-Kluwe K (2015). Climate services for marine applications in Europe. Earth Perspectives Trans disciplinarity Enabled 2015 2:3. doi:10.1186/s40322-015-0029-0

Chapter 6
Organic Matter Mineralization as Reflected in Deep-Water C_T Accumulation

6.1 Total CO_2 Dynamics During Periods of Stagnation and Water Renewal

The deep Gotland Basin is an ideal place to study the kinetics and stoichiometry of biogeochemical transformations. It provides a natural laboratory because during long-lasting periods of stagnation the water masses below 150 m (Fig. 6.1) are almost unaffected by lateral water transport (see Chap. 4). Hence, changes in the concentrations of biogeochemical constituents can be attributed solely to vertical mixing and to the mineralization of POM derived from biomass production in surface waters. The vertical C_T distribution below the permanent halocline was measured on 58 occasions during 2003–2017, hence with an average temporal resolution of $\sim$4 months. Since stagnation periods usually last for many years, this resolution is adequate for investigating the progress and characteristics of mineralization processes. A time series of C_T between 100 m and the sea floor (235 m) at station BY15 (Figs. 4.4 and 6.1) is shown in Fig. 6.2. The Hovmöller diagram is based on a multilevel B-spline approximation (Lee et al. 1997) of C_T between the sampling depth (100, 110, 125, 150, 175, 200, 225, and 235 m) and the sampling date. The C_T at 100 m depth generally ranges between 1900 and 2000 µmol kg^{-1} and is substantially higher than the surface-water C_T, the current value of which is 1600 $\pm$ 50 µmol kg^{-1} (Schneider et al. 2002). The higher C_T below the halocline is a consequence of two effects: (1) The increase in salinity implies an increase in alkalinity, which in turn increases the background C_T. The latter refers to the C_T of the respective water mass at equilibrium with atmospheric CO_2, hence, to a C_T that is not affected by biological activities such as the production or mineralization of OM. According to the relationship between alkalinity and salinity suggested by Müller et al. (2016) for the region between the central Baltic Sea and the Kattegat, which is the source region for the Gotland Sea's deep water, the background C_T changes by $\sim$30 µmol kg^{-1} per salinity (S) unit. Assuming a mean equilibrium C_T of 1600 µmol kg^{-1} in the surface water at S = 7, this means that the background

B. Schneider and J.D. Müller, *Biogeochemical Transformations in the Baltic Sea*, Springer Oceanography, DOI 10.1007/978-3-319-61699-5_6

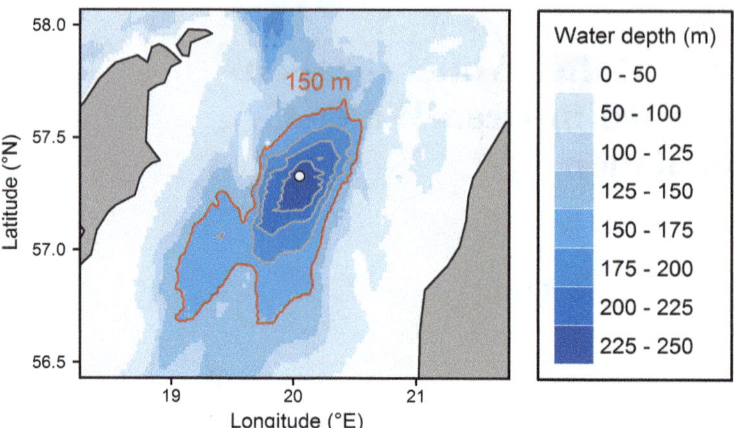

Fig. 6.1 Topography of the Eastern Gotland Basin below 150 m with station BY15 in the centre (white dot)

C_T at 100 m (S = 10.5) and in the bottom water (S = 12.5) amounts to roughly 1700 µmol kg^{-1} and 1770 µmol kg^{-1}, respectively. (2) The second and more important contribution to the deep-water C_T is provided by the mineralization of OM that is supplied either by sinking POM or by lateral particle input. The latter is not necessarily connected with lateral water input into the basin, but results from the interplay between resuspension and sedimentation in connection with horizontal movement of the water masses in and around the basin. The strong vertical C_T gradients that lead to C_T values >2400 µmol kg^{-1} can hence only be explained by intense POM mineralization during a period of stagnation.

The dynamics of POM mineralization as reflected in the accumulation of C_T and the interaction with lateral water exchange are illustrated by the temporal development of the C_T depth-distribution patterns shown in Fig. 6.2. At the start of the C_T measurements, in 2003, the water column was still under the influence of an inflow event that had occurred in 2002. C_T was relatively low and there was no clear vertical structure. Furthermore, concurrent O_2 measurements showed that the entire water column was oxygenated. The situation changed in May 2004, when a new and long-lasting period of stagnation started. C_T enrichment was first observed in the bottom water, which very soon became anoxic. At the same time, the bottom-water C_T diffused upwards and a pelagic interface between the anoxic and oxic waters (Fig. 6.2, black and white points), called the redoxcline, was established that coincided with a C_T level between 2000 and 2100 µmol kg^{-1}. This development was stopped and partly reversed in July 2006, when younger water masses containing less C_T entered the deep water of the Gotland Basin. This did not lead to a re-oxygenation of the Gotland Basin and was, from a hydrological perspective, not considered as a major inflow event in the history of inflows (Mohrholz et al. 2015). Hence, the following increase of C_T in the deep water can be considered as a continuation of the stagnation period that started in May 2004. The

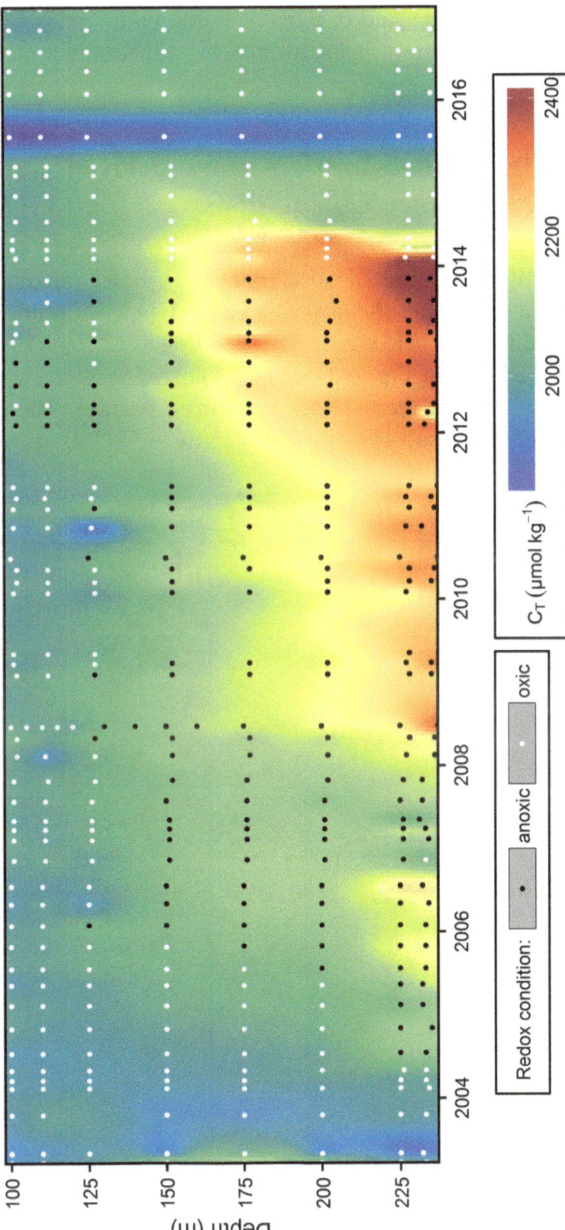

Fig. 6.2 Timeseries of the depth-distribution of total CO_2, C_T, during a stagnation period in the eastern Gotland Basin. *White* and *black dots* indicate occurrence of oxygen and hydrogen sulfide, hence, oxic and anoxic conditions, respectively

propagation of C_T enrichment and thus the migration of the redoxcline towards a depth of ~125 m continued. However, this development ceased abruptly in 2014 and 2015, when the entire deep water of the Gotland Basin was stepwise oxygenated by lateral inflows that peaked in a major Baltic inflow in 2015 (Mohrholz et al. 2015) and marked the starting point for a new stagnation period.

6.2 Organic Matter Mineralization Rates Derived from C_T Mass-Balance Calculations

A previous estimate of OM mineralization rates was based on data obtained between May 2004 and July 2006 (Schneider et al. 2010). During this period, the salinity decreased continuously below a depth of 150 m, as exemplarily illustrated for depths of 150 and 225 m in Fig. 6.3a (Phase 1). This facilitated the determination of mixing coefficients on the basis of a salt mass balance. The calculations performed for each of four sub-layers (150–175 m, 175–200 m, 200–225 m,

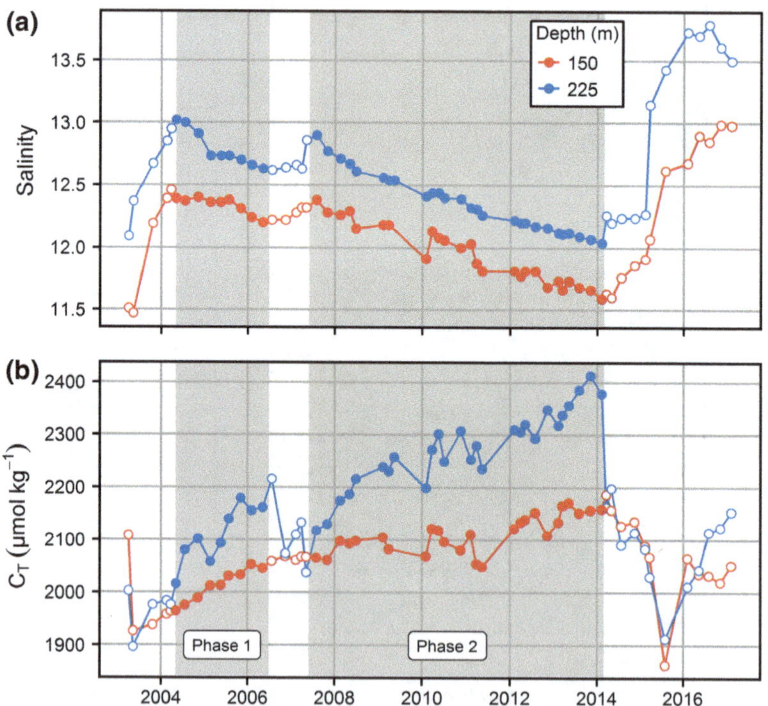

Fig. 6.3 Temporal development of **a** salinity and **b** total CO_2 at depth of 225 and 150 m during the 2004–2014 stagnation period. The shadowed areas (*full circles*) represent two phases of the stagnation which are separated by a transient increase in salinity due to lateral water exchange

225 m–bottom) and for each measurement interval yielded mixing coefficients (Schneider et al. 2010) that agreed well with the results from other studies (Axell 1998; Gustafsson and Stigebrandt 2007).

The expected C_T increase during the stagnation period was irregular and interrupted by occasional phases of decreasing C_T (Fig. 6.3b), especially at the lower depth of 225 m. This could be explained by a mass balance for C_T for the four sub-layers below 150 m, which included vertical C_T exchange by mixing. The amount of C_T actually produced by mineralization was then given by the sum of the observed increases in the C_T concentration (in units of µmol dm^{-3}) and the loss or gain of C_T by mixing between the sub-layers. In contrast to the observed C_T concentration (Fig. 6.3b), the accumulated amount of C_T produced by mineralization increased linearly with time in the different sub-layers below 150 m (Fig. 6.4). The slopes of the regression lines in Fig. 6.4 represent the mineralization rates, expressed in terms of µmol dm^{-3} day^{-1}, for the different sub-layers, which can be converted to the unit mol m^{-2} $year^{-1}$ (right hand scale in Fig. 6.4) by taking into account the volume and the sediment area of the basin. The black dashed line in the figure refers to the mean C_T accumulation in the basin and corresponds to an average annual rate for the basin below 150 m of 2.0 mol m^{-2} yr^{-1} (Schneider et al. 2010).

To examine the representativeness of this relatively short-term rate (Fig. 6.3, Phase 1), in the following similar calculations are presented for the 7-year stagnation period from May 2007 to February 2014 (Fig. 6.3, Phase 2), during which the water column between 150 m and the sea floor was consistently anoxic. The general trend towards lower salinities during this period again indicated that no major inflow event had occurred. While transient increases in salinity of a few hundredths of salinity units might be caused by minor lateral water exchanges, the effect on the mineralization

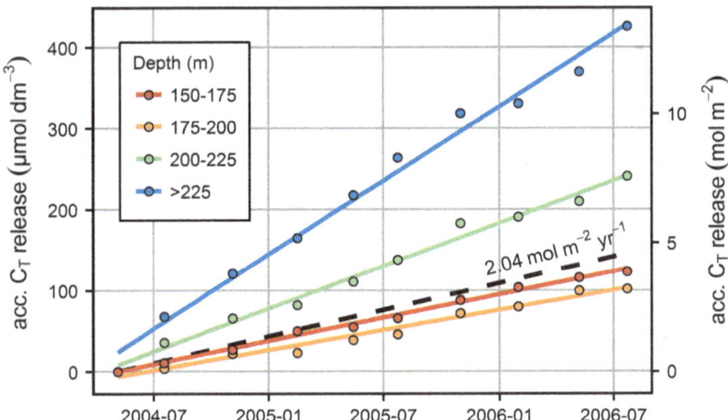

Fig. 6.4 Accumulated C_T release by OM mineralization at different depth level in the Eastern Gotland Basin during the 2004–2006 stagnation phase. The slopes of the regression lines represent the mineralization rates. The *dashed black line* refers to the mean C_T release in the basin below 150 m and yields a mineralization rate of 2.04 mol m^{-2} $year^{-1}$

estimate will be small. In contrast to the above-mentioned calculations, here we use a simpler approach to estimate mineralization that is based on the salinity and C_T mass balance for the entire water mass below 150 m and does not explicitly calculate mixing coefficients. The basic assumptions are: (1) the decrease in the salt inventory below 150 m is equal to the salt flux, F_S, by vertical mixing at the 150-m boundary; (2) the salt flux, F_S, and the C_T flux, F_{CT}, are related to each other by the ratio of the corresponding gradients at 150 m. As shown in Box 6.1, this means (Eq. 6.1):

$$F_{CT} = F_S \cdot \frac{\Delta C_T}{\Delta S} \tag{6.1}$$

and (3) the sum of the increase in the C_T inventory and C_T loss by mixing yields the C_T released by mineralization (details of the computation are given in Box 6.1). The accumulated C_T release calculated in this way shows a clear linear increase during the 7 years of stagnation (Fig. 6.5). The slope of the corresponding regression line represents the daily mineralization rate, which for the period May 2007–February 2014 in the Gotland Basin below 150 m amounted to 0.15 μmol dm^{-3} day^{-1}. Relating this mineralization rate to the sediment surface area yields an annual rate of 1.8 mol m^{-2} yr^{-1}, which is almost identical to the previous estimate for the period May 2004–July 2006 (2.0 mol m^{-2}) (Schneider et al. 2010). This indicates a certain stability of the input of POM to the deep basin and of the mineralization kinetics. The similar mineralization rates also suggest that oxygen conditions do not play a major role in the kinetics of the mineralization, because the deep water below 150 m was partly oxic during the 2004–2006 stagnation phase but anoxic throughout the 7-year period analyzed here. However, this does not necessarily mean that the characteristic rate constants for the POM mineralization are the same in the presence of oxygen and hydrogen sulfide. It is also conceivable that a smaller mineralization rate constant at anoxic conditions led to an accumulation of POM.

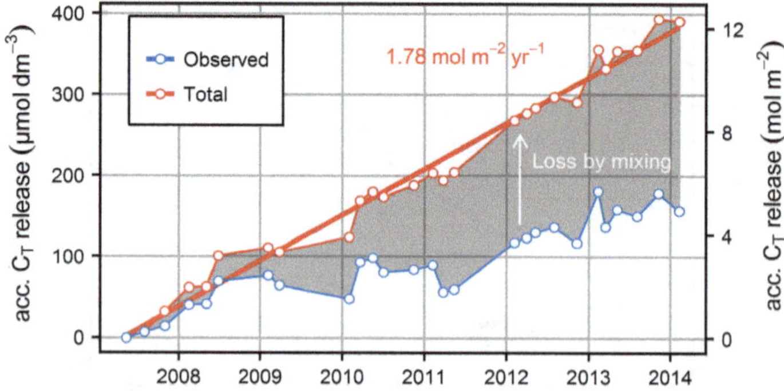

Fig. 6.5 Accumulated C_T release by OM mineralization below 150 m in the Gotland Basin during the 2007–2014 stagnation phase (*red dots* and *line*). The slope of the regression line yields a mineralization rate of 1.78 mol m^{-2} $year^{-1}$. The *blue dots* represent the observed C_T accumulation which together with the C_T loss by vertical mixing at the upper boundary (150 m) account for the mineralization rate

This in turn is increasing the mineralization rate until again a steady state is established during which the mineralization of POM matches the input of POM.

Box 6.1: Mass balance calculations for the determination of mineralization rates

Mass balance calculations that took into account both the observed accumulation of C_T and the C_T loss by vertical mixing were performed for the deep water (below 150 m, in the following called "box") of the Gotland Basin. The calculations were based on C_T and salinity measurements at depth intervals of 25 m between 125 m and 225 m, and in the bottom water at 233 m. The topographic data of Seifert et al. (2001) shown in Table B6.1 were used and C_T and salinity were assumed to have a homogeneous lateral distribution.

The following calculation steps were performed (C_T data originally determined as μmol kg^{-1} were used as if they had the unit μmol dm^{-3}):

1. The basin was subdivided into five different layers according to the measurement depth (Table B6.1). The mean C_T for layers 1–4 (water below 150 m depth), obtained from the values measured at the upper and lower boundaries of each one, was multiplied by the layer volume to obtain the layer C_T inventory. Adding up the inventories of layers 1–4 gave the total inventory for the box below 150 m depth, and dividing by the volume of that compartment the mean box C_T for each measurement point in time.
2. The same procedure was performed for salinity and yielded the mean box salinity, the decrease of which during the stagnation period was attributed to mixing at the upper box boundary. The respective salinity flux (F_S) for each time interval thus corresponds to the differences in the box salinity multiplied by the box volume and divided by the upper box area. Note that F_S always refers to the individual time intervals between measurements and is not normalized to a time unit.

Table B6.1 Topographic data for the Gotland Basin based on the hypsography published by Seifert et al. (2001). Area refers to the upper boundary of the layer

Layer no.	Depth (m)	Area (10^3 km^2)	Volume accumulated (km^3)	Volume layer (km^3)
1	233–225	0.46	3.8	3.76
2	225–200	1.16	24.5	20.8
3	200–175	2.12	64.2	39.7
4	175–150	4.68	149.1	84.9
5	150–125			

3. The C_T flux, F_{CT}, due to mixing at the upper compartment boundary, is given by the mixing coefficient k_{mix} and the C_T gradient at the compartment boundary:

$$F_{CT} = k_{mix} \cdot \frac{\Delta C_T}{\Delta z} \quad \text{(B6.1)}$$

Since the same relationship holds for the mixing of salt:

$$F_S = k_{mix} \cdot \frac{\Delta S}{\Delta z} \quad \text{(B6.2)}$$

F_{CT} can be determined as follows:

$$F_{CT} = F_S \cdot \frac{\Delta C_T}{\Delta S} \quad \text{(B6.3)}$$

where ΔC_T and ΔS refer to the concentration gradient at the upper compartment boundary and are estimated as the difference between the mean values of layers 4 and 5. To calculate F_{CT} for the individual time intervals, the $\Delta C_T/\Delta S$ values at the beginning and end of the intervals were averaged and multiplied by F_S, which was determined according to (2). The F_{CT} value obtained was multiplied by the upper box area and divided by the box volume to obtain the C_T loss in the box attributable to mixing. This value, expressed in terms of concentration units, together with the increase in C_T yielded the amount of C_T released by OM mineralization during each measurement interval.

In a few cases, the mean compartment salinity increased slightly between two successive measurement dates and led to a reversal of the calculated salt and C_T fluxes. In the absence of evidence for a lateral water input, water-column dynamics, such as internal waves, were presumed to have caused the anomalies in the measured salinity trend. Therefore, the salt and C_T fluxes were set to zero for these periods.

Still, there were pronounced differences between the two stagnation phases with regard to the size of the C_T fraction that escaped from the deep basin by vertical mixing at the 150-m boundary. While it was <10% between May 2004 and July 2006, when stagnation was in an early stage after the inflow event in 2003, the situation was entirely different during the long-lasting stagnation period that started in May 2007. This is illustrated by Fig. 6.5, which shows the temporal development of both C_T release and the observed C_T accumulation below 150 m. The difference between these two lines (grey area) represents the C_T fraction that passed the 150-m boundary by vertical mixing. By the end of the 7-year stagnation period, 67% of the

C_T originating from OM mineralization had become mixed into the upper water layers. This shows that it is essential to include vertical mixing when estimating mineralization or any other deep-water process rates on the basis of mass balance. That mixing played only a minor role in the mass balance during the stagnation period (May 2004–July 2006) can be explained by the fact that mineralization takes place mainly at the sediment surface. Therefore, at the start of a stagnation period, strong C_T gradients develop first in deeper water layers of the basin and do not affect C_T exchange across the 150-m boundary. Only with progressing stagnation and the upward migration of the C_T "front," does the removal of C_T from the basin by mixing become increasingly important. This was obviously the case during the 7 years of stagnation.

6.3 Release and Transformations of Nutrients During OM Mineralization

Dissolved silicate (DSi) is not a nutrient in the sense that it contributes to the formation of living OM. Still, the growth of diatoms that form silicate (opal) shells depends on the availability of DSi. We therefore include DSi accumulation in our consideration of nutrients during the May 2007–February 2014 stagnation phase. Silicates forming the hard shells of diatoms are not subjected to mineralization because silicate is a mineral; rather they slowly dissolve, mainly after dead diatoms sink to the seafloor. Unlike released phosphate and ammonia, DSi is not affected by the redox conditions of the water column. Its deep-water accumulation follows that of C_T and during the 7 years of stagnation increased almost linearly with time (Fig. 6.6). Furthermore, the relationship between the DSi that accumulated and the

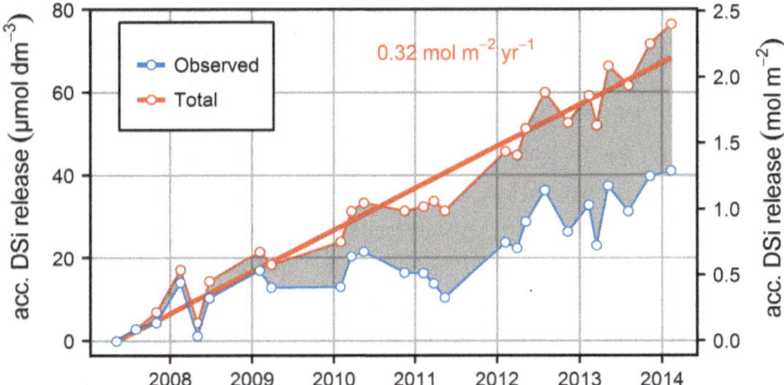

Fig. 6.6 Accumulated dissolution of silicate below 150 m in the Gotland Basin during the 2007–2014 stagnation phase (*red dots* and *line*). The slope of the regression line yields a dissolution rate of 0.32 mol m^{-2} year^{-1}. The *blue dots* represent the observed dissolved silicate accumulation which differ from the total release of silicate because of vertical mixing at the upper boundary (150 m) of the basin

fraction that had passed the 150-m boundary also resembled that of C_T. The mean annual dissolution rate of DSi, as obtained from the slope of the regression line in Fig. 6.6, was 0.32 mol m^{-2} yr^{-1}, that is, ~6-fold lower than the carbon mineralization rate. This ratio roughly corresponds to the C/Si ratio of diatoms that can be inferred from observed N/Si ratios (Brzezinski 1985) and assuming a C/N ratio of 6.6 for the POM of diatoms according to Redfield et al. (1963). However, there are two arguments that challenge the connection between the C/Si ratio in diatoms and the CO_2/DSi ratio measured during deep-water stagnation: First, the biogenic particles that accumulate at the sediment surface are not exclusively produced by a bloom of diatoms but originate from the growth of many other plankton species that do not form silicate shells. Therefore, the ratio of the CO_2/DSi release rates should be shifted towards an enhanced CO_2 release. Second, DSi is generated in deep water by the dissolution of SiO_2 whereas CO_2 is produced by the microbial oxidation of OM. These are two entirely different processes and they will lead to different release rates, because SiO_2 dissolution is slower than POM mineralization (Goutx et al. 2007). Hence, the formal consistency between POM mineralization and SiO_2 dissolution, is somewhat surprising and difficult to explain.

The release of phosphate (PO_4^{3-}) during the mineralization of POM and its successive transformations are of far-reaching importance for Baltic Sea productivity. This is because phosphate may be bound to the insoluble Fe-III-hydroxy-oxides (Fe-III-O-(OH)-PO_4^{3-}) that form in response to oxic conditions but become soluble upon their reduction to Fe(II) in anoxic waters, thus releasing PO_4^{3-}. The interplay between dissolved and particle-bound PO_4^{3-} in the Gotland Basin was analyzed for the May 2004–July 2006 stagnation phase and the preceding anoxic period (Schneider 2011). The PO_4^{3-} that accumulated during anoxic conditions became depleted by a factor of ~3 during an inflow event that transported younger, oxygen-rich and low-nutrient water into the basin. Although the concurrent C_T measurements indicated that ~65% of the decrease in the PO_4^{3-} concentration was caused by dilution, the remaining 35% constituted a significant fraction that was removed from the dissolved phase and obviously precipitated as Fe-III-O-(OH)-PO_4^{3-} at the sediment surface. This process was reversed when the basin again became anoxic at the beginning of the next stagnation period, in May 2004. A sudden release of PO_4^{3-} occurred that could only be explained by the anoxic dissolution of Fe-III-O-(OH)-PO_4^{3-}, because the rate was 2- to 3-fold larger than expected from the mineralization of POM as inferred from the increased C_T. However, the intense PO_4^{3-} release weakened after several months of continuous anoxia and the relationship to the C_T release rates approached the Redfield C/P ratio of 106. This indicated that the Fe-III-O-(OH)-PO_4^{3-} deposits at the sediment surface were exhausted and that the continued PO_4^{3-} release was related to the mineralization of OM. Schneider (2011) concluded from these observations that the high PO_4^{3-} release rates at the start of a stagnation period are a temporary phenomenon that reverse the preceding formation of insoluble Fe-III-O-(OH)-PO_4^{3-} and thus have no net effect on phosphorus availability in the Baltic Sea (Schneider 2011).

The PO_4^{3-} data presented here in connection with C_T measurements refer to the second stagnation phase from May 2007–February 2014. This period can be considered as the continuation of the previous stagnation phase (2004–2006) that was interrupted by a weak water-renewal event. This caused a slight increase in the salinity and a decrease of C_T in the water layers below 150 m (Figs. 6.2 and 6.3a, b), but the concurrent oxygen input was too low to restore the oxic conditions of the deep basin. The accumulated PO_4^{3-} release (Fig. 6.7) again exceeded considerably the observed PO_4^{3-} increase due to mixing across the 150 m boundary. This was especially obvious in the late stage of the stagnation period, when the mean PO_4^{3-} concentrations remained at a roughly constant level. The regression line for the accumulated PO_4^{3-} release as a function of time yielded a mean annual rate of 0.018 μmol m^{-2} whereas the annual release rate for C_T was 1.8 mol m^{-2}. Hence, in accordance with the late phase of the 2004–2006 stagnation, POM mineralization under anoxic conditions behaved quite conservatively and generated C_T and PO_4^{3-} at a ratio close to the classical Redfield ratio.

The first stagnation phase (May 2004–July 2006) was also used to establish a N budget in connection with POM mineralization (Schneider et al. 2010). Similar to the mass balance calculations for C_T, four discrete water layers below a depth of 150 m were considered. However, the aim was not to independently determine the release of ammonium (NH_4^+), as the primary N compound released during mineralization, but to estimate concurrent denitrification. This was achieved based on the imbalance in the N budget, which comprised the observed accumulation of DIN, the vertical mixing of NH_4^+ and NO_3^-, and the release of NH_4^+ by POM mineralization. To quantify the release of NH_4^+ for each time step and sub-layer, a constant ratio between the production of C_T and NH_4^+ was used. This was obtained from a NH_4^+ mass balance calculation for the bottom water and for times when the redoxcline was located above 200 m. Under these circumstances, NH_4^+ concentrations will not be affected by denitrification because the necessary input of oxygen

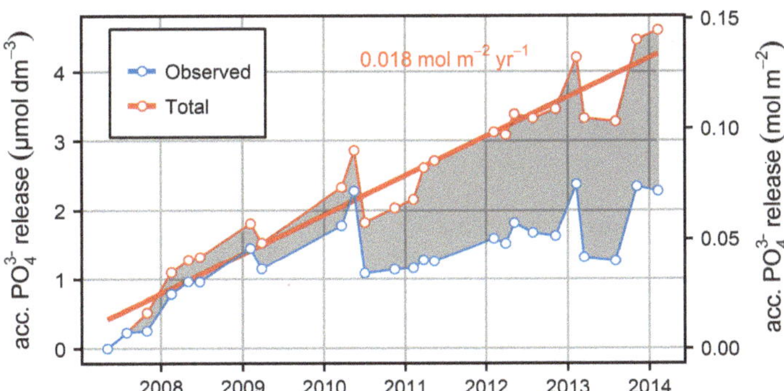

Fig. 6.7 Accumulated release of phosphate by OM mineralization below 150 m in the Gotland Basin during the 2007–2014 stagnation phase (*red dots* and *line*). The slope of the regression line yields a release rate of 0.018 mol m^{-2} $year^{-1}$. The *blue dots* represent the observed PO_4^{3-} accumulation which differ from the total release of PO_4^{3-} because of mixing at the upper boundary (150 m) of the basin

or nitrate by mixing can be excluded. Relating the resulting NH_4^+ release to the concurrently determined C_T production yielded a C/N ratio of 8.8. This ratio was then used together with the amount of C_T release to determine the amount of NH_4^+ produced by mineralization in the different water layers and for different time intervals during the stagnation period. Including this as a source term into the mass balance calculations yielded the loss of DIN (ammonium + nitrate) attributable to denitrification. The denitrification obtained in this way amounted to 0.56 mol m^{-2} for the entire considered stagnation period (27 months). However, mass balance indicated that denitrification largely stopped after 15 months, when the redoxcline had passed over the 150-m boundary, hence, when the basin had become entirely anoxic. Therefore, it is not meaningful to normalize the estimated denitrification to a time unit, e.g., per year, but to refer it to the transition of the basin from oxic to anoxic conditions. According to these findings, our simplified DIN mass balance for the long-lasting stagnation phase from May 2007 to July 2014 did not yield significant denitrification, because the entire basin was anoxic right from the start.

References

Axell LB (1998) On the variability of Baltic Sea deepwater mixing. J Geophys Res 103 (C10):21667–21682

Brzezinski MA (1985) The Si:C: N ratio of marine diatoms: interspecific variability and the effect of some environmental variables. J Phycol 21:347–357

Goutx M, WakehamS G, Lee C, Duflos M, Guigue C, Liu Z, la Moriceau B, Sempe R, Tedetti M, Xue J (2007) Composition and degradation of marine particles with different settling velocities in the northwestern Mediterranean Sea. Limnol Oceanogr 52(4):1645–1664

Gustafsson BG, Stigebrandt A (2007) Dynamics of nutrients and oxygen/hydrogen sulphide in the Baltic Sea deep water. J. Geophys. Res. 112 (C10): 21,667–21,682

Lee S, Wolberg G, Shin SY (1997) Scattered data interpolation with multilevel B-splines. IEEE transactions on Trans Visual Comput Graphics 3(3):228–244

Mohrholz V, Naumann M, Nausch G, Krüger S, Gräwe U (2015) Fresh oxygen for the Baltic Sea—An exceptional saline inflow after a decade of stagnation. J. Mar. Sys. 148:152–166

Müller JD, Schneider B, Rehder G (2016) Long-term alkalinity trends in the Baltic Sea and their implications for CO_2-induced acidification. Limnol. Oceanogr. 61:1984–2002. doi:10.1002/lno.10349—open access

Redfield AC, Ketchum BH, Richards FA (1963). The influence of organisms on the composition of sea water. In: Hill MN (ed), The Sea. Interscience, vol 2. New York, pp. 26–77

Schneider B (2011) PO_4 release at the sediment surface under anoxic conditions: a contribution to the eutrophication of the Baltic Sea? Oceanologia 53:415–429

Schneider B, Nausch G, Kubsch H, Petersohn I (2002). Accumulation of total CO_2 in the Baltic Sea deep water and its relationship to nutrient and oxygen concentrations. Mar Chem 77:277–291

Schneider B, Nausch G, Pohl C (2010). Mineralization of organic matter and nitrogen transformations in the Gotland Sea deep water. Mar. Chem. 119:153–161

Seifert T, Tauber F, Kayser B (2001) A high resolution spherical grid topography of the Baltic Sea. Baltic Sea Science Congress, Stockholm. Abstract volume, Stockholm Marine Research Centre, Stockholm University.

Chapter 7
Progress Made by Investigations of the CO_2 System and Open Questions

In the introduction we stated that investigations of the marine CO_2 system are an ideal tool for biogeochemical studies because almost all biogeochemical processes are ultimately driven by the interplay between organic matter (OM) production and mineralization in an everlasting cycle driven by solar radiation. Since these processes are connected with the consumption or production of CO_2, CO_2 mass balance calculations based on measured changes in total CO_2, were successfully used to estimate net production and mineralization rates in the Baltic Sea. The utility of this approach was demonstrated in Chaps. 5 and 6, which discussed the dynamics and control of both net community production (NCP) in the surface water and OM mineralization in stagnant deep basins, respectively. These types of studies play a central role in Baltic Sea research because they are fundamental to an understanding of ecosystem functioning and to assessments of eutrophication and its consequences.

Regarding OM production studies, the advantages of our approach are amplified by the use of an automated pCO_2 measurement system deployed on a cargo ship that traverses the entire Baltic Proper up to five times per week. Using this system, we obtained pCO_2 data with a resolution of 1–2 days. This allowed us to resolve the processes which occur during major production events and usually last not much longer than 2 weeks.

By this combination of a new scientific approach and innovative technology, we were able to expand our knowledge and understanding of the production processes in the Baltic Sea. More specifically, the major findings are:

- The seasonality of the pCO_2 in the Baltic Proper and in the western part of the Gulf of Finland is characterized by two distinct minima that reflect the major NCP periods: the spring bloom and mid-summer production fueled by N-fixation.
- The start of the spring bloom is not related to a temperature threshold but coincides with an increase in temperature. Except in the southwest Baltic Sea, between 2004 and 2014 the spring bloom consistently started by the end of March, within a time slot of only a few days.

B. Schneider and J.D. Müller, *Biogeochemical Transformations in the Baltic Sea*, Springer Oceanography, DOI 10.1007/978-3-319-61699-5_7

- In most years spring bloom NCP continues after the complete exhaustion of nitrate that occurs regularly by mid-April. Calculation of a total N budget suggested that N fixation takes place already in spring and provides the N needed for NCP. However, there is as yet no evidence for the early existence of N-fixing organisms in the Baltic Proper and the N source used in the post-nitrate bloom remains unknown.
- Mid-summer NCP fueled by N-fixation and identifiable as a drop in the total CO_2 in surface water takes place intermittently. Several NCP pulses may occur and they coincide with pulses of a strong increase in temperature caused by the low vertical mixing during calm weather conditions. The accumulated NCP for the individual pulses examined in this and earlier studies correlated linearly with temperature in all cases. Since the temperature increase results from the energy input by solar radiation, it was concluded that also N-fixation and the connected NCP were controlled by the exposition to solar radiation.
- In each of the observed cases, the simultaneous increases in the accumulated mid-summer NCP and in temperature continued as long as low wind conditions persisted. This suggests that mid-summer N-fixation and related NCP is only controlled by the mean exposure to radiation which results from the radiation intensity divided by the mixed layer depth, and that the availability of phosphorus is not a limiting factor.

However, at the same time these findings raise new questions. In our view, there are two major issues that require scientific attention and research efforts. The first is related to the continuation of the NCP after the complete nitrate depletion in early spring. Indications for the occurrence of early N-fixation in spring which were based on mass balance calculations, could not be confirmed by N-fixation rate measurements. Hence, the nitrogen supply for the second phase of the spring bloom remains an open question. Since the spring NCP after the nitrate depletion is approximately equivalent to the nitrate fuelled NCP, the identification of the respective nitrogen source is of great importance for the assessment of the Baltic Sea eutrophication either by monitoring or model simulations.

The second unsolved question that emerged from our investigations is related to the mid-summer NCP fuelled by N-fixation. POM production continued in the absence of phosphate and accordingly extremely high C/P ratios were observed in the produced POM. This raises the question for the nature of the POM. Since the C/N ratios in POM did not show any anomalies (7.5) compared to POM produced during the spring bloom, it was speculated that in the absence of phosphate N-fixation by cyanobacteria was used for the production of solely proteins, possibly as extracellular material. However, this is a pure speculation.

Beyond the research needs for deepening our biogeochemical process understanding, we would like to emphasize the high potential of surface water pCO_2 measurements to detect long-term NCP trends. This is a central task of the eutrophication monitoring executed by the Baltic Sea riparian countries. In contrast to traditional measurements of nutrient and/or chlorophyll *a* concentrations, high resolution pCO_2 records facilitate more reliable NCP estimates and are thus more sensitive to the detection of trends.

We could also demonstrate that measurements of C_T in stagnant deep-water layers are an ideal tool to study OM mineralization. This was shown by our long-term observations of the C_T accumulation in the Gotland Sea, at water depths below 150 m. The major findings, which emerged from the analysis of a stagnation period that lasted from May 2004–February 2014, are:

- Vertical mixing plays a key role in the accumulation of mineralization products (C_T, nutrients) and becomes increasingly important with the increasing duration of the stagnation period. It must therefore be taken into account in mass balance calculations.
- The annual mineralization rates determined independently for two phases of the long-lasting stagnation period were 1.8 mol m^{-2} $year^{-1}$ (2004–2006) and 2.0 mol m^{-2} $year^{-1}$ (2007–2014), respectively. A dependency of the mineralization rates on the redox conditions, which changed from oxic to anoxic during the first phase of the stagnation, could not be detected.
- The enhanced release of phosphate at the beginning of the stagnation period in 2004 could be explained by the anoxic dissolution of the Fe-III-O-(OH)-PO_4^{3-} that precipitated during the preceding inflow event. Hence, the high phosphate release rate was simply the reversal of the preceding removal and had no net effect on the phosphate budget. After the entire basin below 150 m had become anoxic, the release of phosphate decreased and, when measured as a function of the release of C_T, yielded a C/P ratio that approached the Redfield ratio.
- Nitrogen mass balance calculation revealed the amount of denitrification that occurred during the transition of the considered basin from oxic to entirely anoxic. Denitrification that occurred during that 15-month observation period was ~560 mmol-N m^{-2}, approximately four times more than the annual nitrogen input by N-fixation. However, this comparison does not allow a determination of the balance between denitrification and N fixation because denitrification will continue after the redoxcline rises above the 150-m that was the upper boundary for our mass balance calculations.

The determination of the annual POM mineralization rate and its relationship to the release of nutrients are of importance for the entire central Baltic Sea. This is because the Gotland Basin acts as an accumulation area where POM produced in the surface water of a much larger area is collected and finally mineralized at the sediment surface. In contrast to the general belief, our data suggest that the mineralization rate does not depend on the redox conditions. However, this does not necessarily imply that the mineralization rate constants are identical. For a deeper understanding and parameterization of the POM mineralization process, we therefore need an improved characterization of the mineralization rate constants. This is a challenge for further deep water CO_2 studies which necessarily imply investigations of the vertical and lateral input of POM and its burial in the sediment, hence, a comprehensive carbon mass balance.

Index

B. Schneider and J.D. Müller, *Biogeochemical Transformations in the Baltic Sea*, Springer Oceanography, DOI 10.1007/978-3-319-61699-5

Zeitfracht Medien GmbH
Ferdinand-Jühlke-Straße 7
99095 Erfurt, Deutschland
produktsicherheit@kolibri360.de